Praise for

THE BIRTH OF THE PILL

"Nimbly paced and conversational. . . . Eig . . . brings a lively, jocular approach to the story."

—Irin Carmon, *The New York Times Book Review*

"Dynamic, highly engrossing. . . . As hard as it is to put down *The Birth of the Pill*'s story of four privileged individuals' thrilling quest for better living through science, it's imperative to remember the scores of women lost to history whose flesh and blood helped make it a reality."

—Anna Holmes, *The Los Angeles Times*

"A fresh, infectiously readable take."

—Kate Tuttle, *The Boston Globe*

"Masterful. . . . [W]hen legislatures and courts threaten to negate the miracles of science and human progress so dazzlingly portrayed here, Eig's book is essential reading."

—Kate Manning, *The Washington Post*

"A tale of scientific progress and social change as engaging and gripping as any suspense novel. . . . Prescient."

—*Elle*

"Excellent."

—*The Chicago Sun-Times*

"A gripping story of scientific discovery and a reminder that women's emancipation is a recent and hard-won victory . . . a fitting tribute to their struggle."

—Isabel Berwick, *The Financial Times*

"Well-crafted and comprehensive."

—Miriam Krule, *Slate*

"Surprisingly gripping."

—*The Chicago Tribune*

"An astonishing story, told with detail and even excitement."

—Rob Hardy, *The Columbus Dispatch*

"Vivid and life-affirming."
—*The Guardian*

"The pill is utterly ordinary today. The story of how we got here is anything but."
—Hannah Levintova, *Mother Jones*

"Eig's research is thorough and his account exhaustive."
—Emily Witt, *Bookforum*

"Evocative and informative."
—*Health Affairs*

"Superbly written and well documented . . . delightful, thought-provoking reading."
—Amy Ong Tsui, *Studies in Family Planning*

"Riveting. . . . Written with pace and clarity, *The Birth of the Pill* is a vivid portrait of four brilliant and courageous misfits."
—*The Telegraph*

"Vivid . . . a rebuke to all those who lambast the Pill for unleashing promiscuity. . . . [Eig] also has an eye for the wider social canvas."
—*The Times* [London]

"Rousing and involving."
—*The Independent*

"A pacy narrative, full of sharp characterisation."
—*The Sunday Times* [London]

"A fast-paced . . . thought-provoking social history. Weaving medical, corporate, and political history with rich biographical detail, Eig turns the history of the pill into a scientific suspense story full of profoundly human characters. The result is cultural history at its finest."
—Alan W. Petrucelli, *The Examiner*

"A page-turning, suspense-filled, beautifully written narrative about the history of the pill . . . an irresistible tale."
—Ken Burns

"'The pill' is that rare invention that transforms the world. In this gripping book, Jonathan Eig tells how an unlikely group—Margaret Sanger, Katherine McCormick, Dr. Gregory Pincus, and Dr. John Rock—came together to achieve a scientific breakthrough and win acceptance for it in the face of intense opposition. The Birth of the Pill is vivid, compelling, and important."

—T. J. Stiles, Pulitzer Prize–winning author of *The First Tycoon: The Epic Life of Cornelius Vanderbilt*

"Jonathan Eig turns the history of the pill into a smart and spicy account of the unlikely bonds that linked a millionaire activist, a free-loving crusader, a Roman Catholic gynecologist, and a maverick scientist. *The Birth of the Pill* is at once intelligent, well researched, witty, and captivating. . . . [A] unique prism into the changing morals about sex, women, and marriage in twentieth-century America."

—Randi Hutter Epstein, M.D., author of *Get Me Out: A History of Childbirth from the Garden of Eden to the Sperm Bank*

"'The pill' is that rare invention that transforms the world. In this gripping book, Jonathan Eig tells how an unlikely group—Margaret Sanger, Katherine McCormick, Dr. Gregory Pincus, and Dr. John Rock—came together to achieve a scientific breakthrough and win acceptance for it in the face of intense opposition. *The Birth of the Pill* is vivid, compelling, and important."

—T. J. Stiles, Pulitzer Prize–winning author of *The First Tycoon: The Epic Life of Cornelius Vanderbilt*

"Jonathan Eig turns the history of the pill into a smart and spicy account of the unlikely bonds that linked a millionaire activist, a free-loving crusader, a Roman Catholic gynecologist, and a maverick scientist. *The Birth of the Pill* is at once intelligent, well researched, witty, and captivating. . . . [A] unique prism into the changing morals about sex, women, and marriage in twentieth-century America."

—Randi Hutter Epstein, M.D., author of *Get Me Out: A History of Childbirth from the Garden of Eden to the Sperm Bank*

THE BIRTH *of* THE PILL

ALSO BY JONATHAN EIG

Ali:
A Life

Get Capone:
The Secret Plot That Captured America's
Most Wanted Gangster

Opening Day:
The Story of Jackie Robinson's First Season

Luckiest Man:
The Life and Death of Lou Gehrig

THE BIRTH *of* THE PILL

How Four Crusaders Reinvented Sex and Launched a Revolution

JONATHAN EIG

W. W. NORTON & COMPANY
Independent Publishers Since 1923
NEW YORK | LONDON

Excerpt of "Annus Mirabilis" from *The Complete Poems* by
Philip Larkin © 2012, reprinted by permission of Faber & Faber Ltd.

Printed in the United States of America
First published as a Norton paperback 2015

For information about special discounts for bulk purchases, please contact
W. W. Norton Special Sales at specialsales@wwnorton.com
or 800-233-4830

Manufacturing by LSC Harrisonburg
Book design by Dana Sloan
Production manager: Devon Zahn

Library of Congress Cataloging-in-Publication Data

Eig, Jonathan.
The birth of the pill : how four crusaders reinvented sex and launched a
revolution / Jonathan Eig.
pages cm
Includes bibliographical references and index.
ISBN 978-0-393-07372-0 (hardcover)
1. Oral contraceptives—History. 2. McCormick, Katharine Dexter,
1876–1967. 3. Pincus, Gregory, 1903-1967. 4. Rock, John, 1890-1984
5. Sanger, Margaret, 1879-1966. I. Title.
RG137.5.E34 2014
618.1'82209—dc23

2014019355

ISBN 978-0-393-35189-7 pbk.

W. W. Norton & Company, Inc.
500 Fifth Avenue, New York, N.Y. 10110
www.wwnorton.com

W. W. Norton & Company Ltd.
15 Carlisle Street, London W1D 3BS

4 5 6 7 8 9 0

For Jennifer

Sexual intercourse began
In nineteen sixty-three
(which was rather late for me)—
Between the end of the Chatterley ban
And the Beatles' first LP.

—PHILIP LARKIN, "ANNUS MIRABILIS"

Contents

THE BIRTH *of* THE PILL

ONE

A Winter Night

Manhattan, Winter 1950

She was an old woman who loved sex and she had spent forty years seeking a way to make it better. Though her red hair had gone gray and her heart was failing, she had not given up. Her desire, she said, was as strong and simple as ever: She wanted a scientific method of birth control, something magical that would permit a woman to have sex as often as she liked without becoming pregnant. It struck her as a reasonable wish, yet through the years one scientist after another had told her no, it couldn't be done. Now her time was running out, which was why she had come to an apartment high above Park Avenue to meet a man who was possibly her last hope.

The woman was Margaret Sanger, one of the legendary crusaders of the twentieth century. The man was Gregory Goodwin Pincus, a scientist with a genius IQ and a dubious reputation.

Pincus was forty-seven years old, five feet ten and a half inches tall, with a bristly mustache and graying hair that shot from his head in every direction. He looked like a cross between Albert Einstein and Groucho Marx. He would speed into a room, working a Viceroy between his yellowed fingers, and people would huddle close to hear what he had to say. He wasn't famous. He owned no scientific prizes. No world-changing inventions were filed under his name. In fact, for a long stretch of his career he had been an outcast from the scientific establishment, rejected as a radical by Harvard, humiliated in the press, and left with no choice but to conduct his varied and oftentimes controversial experiments in a converted garage. Yet he radiated confidence as if he knew the world would one day recognize his brilliance.

Pincus was a biologist and perhaps the world's leading expert in mammalian reproduction. In the 1930s, at the start of his professional career, he'd attempted to breed rabbits in Petri dishes using much the same technology that decades later would lead to *in vitro* fertilization for humans. Then he was young and handsome and possessed of a limitless imagination. He posed for newspaper photographs and boasted to reporters that a new age of human reproduction was on the horizon, one in which men and women soon would employ modern methods to control the process of making babies. Science would lead the way.

But Americans were not ready to hear such things. The press compared him to Victor Frankenstein, Mary Shelley's fictional scientist, who tried to conjure life but accidentally created a monster. Harvard denied Pincus tenure, and no other university would hire him. He was deemed too dangerous.

At that point, a more humble man might have chosen a new line of work. A weaker man might have succumbed to anger or despair. But not Goody, as his friends and family called him, as much for his friendly nature as his middle name. For while Pincus was affectionate and disarming in social settings, when it came to his career he was, as one colleague put it, "a street-fighting Jew." Getting knocked down was merely the thing that happened before Pincus got up to fight again. When Harvard dumped him and no other job offers arrived, he moved to Worcester, Massachusetts, a factory town, where a former colleague from Harvard had offered him a low-paying, low-ranking position as a researcher for Clark University. He worked in a basement lab where dust from a nearby coal bin contaminated his experiments. When he asked the university to provide him a proper laboratory, the request was denied.

Again, he might have quit. Instead, Pincus and one of his colleagues, Hudson Hoagland, did something unprecedented: they launched their own scientific research center. They went door to door in Worcester (pronounced *wuhstah*, in the local tongue) and the surrounding area, distributing brochures and asking housewives, plumbers, and hardware store owners to contribute—no donation too small—to a new institution they called the Worcester Foundation for Experimental Biology. With the money they scraped together, they bought an old house in nearby Shrewsbury, where Pincus set up his office and lab in the garage. The operation was so lean in those early years that he cleaned his own animal cages and, at one even lower moment, moved his wife and children into a state-run insane asylum while conducting research there on schizophrenia.

※

Pincus knew about Sanger. Almost everyone in America did. It was Sanger who had popularized the term "birth control" and almost single-handedly launched the movement for contraceptive rights in the United States. Women would never gain equality, she had argued, until they were freed from sexual servitude. Sanger had opened the nation's first birth control clinic in Brooklyn in 1916 and helped launch dozens more around the world. But even after decades of work, the contraceptive devices available at those clinics—condoms and cervical caps, mostly—remained ineffective, impractical, or difficult to obtain. It was as if she'd been teaching starving people about nutrition without giving them anything healthy to eat. Sanger explained to Pincus that she was looking for an inexpensive, easy-to-use, and completely foolproof method of contraception, preferably a pill. It should be something biological, she said, something a woman could swallow every morning with her orange juice or while brushing her teeth, with or without the consent of the man with whom she was sleeping; something that would make sexual intercourse spontaneous, with no forethought or messy fumbling, no sacrifice of pleasure; something that would not affect a woman's fertility if she wished to have children later in life; something that would work everywhere from the slums of New York to the jungles of southeast Asia; something 100 percent effective.

Could it be done?

The other scientists she'd approached, every one of them, had said no, and they had given her a long list of reasons. It was dirty, disreputable work. The technology wasn't there. And even if it somehow could be done, there would be no point. Thirty states and the federal government still had anti-birth-control laws on the books. Why go to the trouble of making a pill no drug company would dare to manufacture and no doctor would dare prescribe?

But Sanger held out hope that Gregory Pincus was different, that he might be bold enough—or desperate enough—to try.

※

It was the midpoint of the century. Scientists were taking up matters of life and death that once had been the domain principally of artists and philosophers. Men in lab coats—and yes, they were almost all men—were heroes, winners of wars, battlers of disease, givers of life. Malaria, tuberculosis, and syphilis were among the many illnesses surrendering to modern medicine. Governments and giant corporations poured unprecedented sums of money into research, sponsoring everything from high school science clubs to cold fusion exploration. Health became a political issue as well as a social one. World War II had scarred the earth but also transformed it, offering the promise of a better, freer world, and scientists were leading the way.

Americans were settling into new suburban box homes and exploring the joys of lawn care, dry martinis, and *I Love Lucy*. At least to the casual observer, the United States in the early 1950s appeared staid and steadfast. The Andrews Sisters sang "I Wanna Be Loved" and John Wayne starred in *Sands of Iwo Jima*, celebrating the nation's military might and commitment to democratic ideals.

It was a glorious time to be an American. Young men returning from battle were looking for new adventures and new ways to feel like heroes as they adjusted to the dullness of their homes, marriages, and jobs. During the war, new rules of morality had applied. Sex had become a more casual endeavor as foreign women traded their bodies to American soldiers for cigarettes and cash. Girlfriends back home had written steamy letters filled with promises of the great passion awaiting their men. In truth, many of the women back home had been exploring their own new moral standards. The war had thrust women into the workplace, putting money in their pockets and liberating them from their parental homes. They'd begun dating and making love to men they did not intend to marry, experimenting with new ideas about intimacy and commitment. In 1948, a college professor in Indiana named Alfred Charles Kinsey published a study called *Sexual Behavior in the Human Male*, to be followed five years later by *Sexual Behavior in the Human Female*, and found that people were much friskier than they cared to admit, with 85 percent confessing to premarital sex, 50 percent acknowledging extramarital affairs, and almost everyone saying they masturbated. It would turn out that Kinsey was perhaps biased in his conclusions, but the impact of his work was nevertheless profound. In 1949, Hugh Hefner, a graduate student in sociology at Northwestern University, read Kinsey's report and wrote a term paper arguing for an end to the repression of sex and sexuality in America. "Let us see if we cannot begin to find our way out of this dark, emotional, taboo-ridden labyrinth and into the fresh air and light of reason," Hefner wrote, as he began preparing to do something about it personally.

Late one winter night in Manhattan, Margaret Sanger met Gregory Pincus to talk about nothing less than a revolution. No guns or bombs would be involved—only sex, the more the better. Sex without marriage. Sex without children. Sex redesigned, re-engineered, made safe, made limitless, for the pleasure of women.

Sex for the pleasure of women? To many, that idea was as unthinkable in 1950 as putting a man on the moon or playing baseball on plastic grass. Worse, it was dangerous. What would happen to the institutions of marriage and family? What would happen to love? If women had the power to control their own bodies, if they had the ability to choose when and whether they got pregnant, what would they want next? Two thousand years of Christianity and three hundred years of American Puritanism would come undone in an explosion of uncontrollable desire. Marriage vows would lose their meaning. The rules and roles of gender would be revocable.

Science would do what the law so far had not; it would give women the chance to become equal partners with men. This was the technology Sanger had been seeking all her life.

So, in a sleek Park Avenue apartment where long threads of cigarette smoke floated toward the ceiling, Sanger gazed across a coffee table at Pincus and made her pitch. She was seventy-one years old. She needed this. So did he.

"Do you think that it would be possible . . . ?" she asked.

"I think so," Pincus said.

It would require a good deal of research, he added, but, yes, it was possible. Sanger had been waiting much of her life to hear those words.

"Well," she said, "then start right away."

※

The next morning, Pincus gunned the engine on his Chevrolet, snaking in and out of traffic toward Massachusetts as Sanger's plea snaked in and out of his overactive brain. Driving was new to him. He had only recently inherited this, his first car, from a scientist who had moved abroad, and he was thrilled to discover the speed and power at his command. Driving, like so much else in his life, became a competitive sport. His passengers would white-knuckle their armrests and ask why he was in such a hurry, but Pincus, utterly calm behind the wheel, thought little of it. "This is just my cruising speed," he would say.

The 180-mile journey was full of stops and starts. Interstate highways were yet to come; for now, there were narrow, two-lane roads with slow-downs for school zones and train tracks. The long drive through cold, gray towns and hibernating farm plots gave Pincus time to reflect on his meeting with Sanger.

For as long as men and women have been making babies they've also been trying not to. The ancient Egyptians made vaginal plugs out of crocodile dung. Aristotle recommended cedar oil and frankincense as spermicides. Casanova prescribed the use of half a lemon as a cervical cap. The most popular and effective form of birth control in the early 1950s was the condom, a simple device that dated to the mid-1500s when the Italian doctor Gabriele Falloppio tested a "linen cloth made to fit the glans" to prevent the spread of syphilis. Since Falloppio, though, not much had changed. Condoms became cheaper and more widely available when the Goodyear company began vulcanizing rubber in the 1840s. Crudely fitted cervical caps—an early form of the diaphragm—began to appear at roughly the same time. But in the century that followed, little thought and even less effort had gone into innovation in the field. Pincus had no interest in those antiquated approaches. In his mind, inventing a birth-control pill—inventing anything, for that matter—did not have to be complicated. It was like driving. Step one: Choose your destination. Step two: Select a route. Step three: Try to get there as quickly as possible.

Instead of heading home, he drove to his office at the Worcester Foundation for Experimental Biology to speak with one of his researchers, M. C. Chang. By 1950, Pincus and Hoagland had moved the Foundation from a renovated barn in Worcester to an ivy-covered brick home in a residential section of nearby Shrewsbury. "Outsiders have sometimes called the two-story Foundation building 'the old ladies' home,'" noted the *Worcester Telegram*. "That's what it looks like from the Boston Post Road which runs by the door."

Pincus and Hoagland did their best to make the old ladies' home look like a hall of science. They converted the sun porch to a library. Bedrooms became laboratories. One bedroom-turned-laboratory became a bedroom again when Chang arrived from China by way of Scotland and England to work with Pincus. Though Chang spoke little English, Pincus had spotted something in the scientist, enticing him to join the Foundation for the paltry salary of $2,000 a year (or about $26,000 by today's standards). Chang, who knew Pincus by reputation, thought he would be working in one of America's prestigious institutes and that his fellowship would include free lodging, perhaps on campus, or at least nearby. He did get free lodging, but his room was at the YMCA. He and Pincus would travel to and from work by bus. Later, he would move to the Foundation, sleeping on a small bed in the corner of a converted laboratory and using Bunsen burners to heat his meager meals. As a strict Confucian, Chang didn't mind. He reported proudly that for one important experiment in 1947 he had stored fertilized rabbit eggs in his kitchen refrigerator.

Pincus told Chang that he had spoken to Margaret Sanger about her desire for a pill to prevent pregnancy. It had to be a pill, he explained, not an injection, jelly, liquid, or foam, and not a mechanical device used in the vagina. When Pincus talked in this way—with a sense of purpose, hands chopping at the air, his eyes glittering beneath those bushy brows—his colleagues paid attention.

Goody Pincus was not one of those soft-spoken geniuses content to let his work speak for itself. He was a powerfully built man with a lean, muscular frame. Though his suits and ties were invariably cheap and occasionally mismatched, he nevertheless carried himself with aristocratic self-possession. His voice was stentorian. Confidence was one of his strongest tools. He understood something many scientists did not: that scientific exploration and experimentation were only parts of the job; another equally important part was selling. An idea, no matter how good, might easily die if it were not aggressively pitched—to other scientists, to backers with deep pockets, and, ultimately, to the public. It was the selling that had helped sink him at Harvard, but Pincus was undeterred. He knew from the start that it would be one thing to build a birth-control pill and another to persuade the world to accept it. The scientist attempting such a task would have to be prepared to do both, or there would be little point trying.

Pincus and Chang discussed a scientific paper from 1937—"The Effect of Progesterin and Progesterone on Ovulation in the Rabbit," by A. W. Makepeace, G. L. Weinstein, and M. H. Friedman of the University of Pennsylvania. It reported that injections of the hormone progesterone prevented ovulation in rabbits. Though it had been a huge discovery at the time, no one had tried to explore the implications for humans. There were many reasons. For one thing, scientists weren't seeking innovations in contraception. There was neither prestige nor money in the work, only risk. And even if they had tried, progesterone was too expensive at that time to be widely used.

But when Pincus met Sanger and listened to her plea, attitudes on birth control were shifting—at least a little. Perhaps more important, however, was the evolution then taking place in the field of biology. Scientists were beginning to understand the inner workings of the body well enough to tinker with them. Before the 1950s drugs were mostly developed with the "suck-and-see" approach, as the British referred to trial-and-error experiments. A scientist would concoct a formula in a lab, gulp it down like Dr. Jekyll, and see what effects it had. But those days were nearing an end. Pincus and Chang knew how progesterone functioned. Now the task was to see if they could produce it, modify it, and put it to use. Fortunately, new technology was making progesterone less expensive to obtain. If Sanger would pay for it, Pincus thought he had a good idea of how to proceed.

Pincus was no mere scientific technologist. He had the soul of a romantic. He looked to nature not only for answers but also for beauty. And here was something beautiful. Between puberty and menopause, women normally produce an egg roughly every twenty-eight days from one of their ovaries. The egg migrates down the fallopian tube to the uterus. If the woman has sex with a man and the man ejaculates, five hundred million sperm fight to fertilize her egg. If the egg is not fertilized, it can't implant itself in the lining of the womb, and if it can't implant itself, it is discharged along with the lining of the uterus. If it is fertilized, after about six days the egg can attach to the wall of the uterus, where the woman's blood will nourish it through the placenta. During this gestation, pregnancy begins: A zygote becomes an embryo and an embryo becomes a fetus. Two sex hormones, estrogen and progesterone, guide this process. Pincus focused largely on progesterone.

Often referred to as the pregnancy hormone, progesterone regulates the condition of the inner lining of the uterus. When an egg is fertilized, progesterone prepares the uterus for implantation and shuts down the ovaries so no more eggs are released. In effect, Pincus recognized, nature already had an effective contraceptive. Progesterone was preventing further ovulation to allow the fertilized egg to grow safe from harm. What if the same contraceptive could be delivered in a tablet form, effectively tricking the woman's body into thinking that it was already pregnant? A woman would be able to shut down ovulation any time she liked for as long as she liked. If she didn't release eggs, she couldn't become pregnant.

To Pincus, it was a solution elegant in its simplicity. It wasn't new. It wasn't radical. It was merely a matter of thinking differently about how to solve a problem.

He and Chang began by repeating the experiment done in Pennsylvania, adjusting the dosages and means of delivery to get a feel for progesterone and how it worked. They started with rabbits. Pincus sent a request for funding to the Planned Parenthood Federation of America, the women's health and advocacy group that Sanger had helped found. He asked for $3,100: a $1,000 stipend for Chang, $1,200 for the purchase of rabbits, $600 for animal food, and $300 for miscellaneous supplies.

"I have $2,000, perhaps a little more," Sanger wrote to Pincus a few weeks after their meeting. "Will this do?"

"The amount was ludicrous," Pincus recalled, "but I at once replied, 'Yes.'"

TWO

A Short History of Sex

FOR ALL ITS emotional resonance, not to mention its essential role in the survival of the human species, sex was a subject seldom studied in science.

In the 1950s, William Masters and Virginia Johnson observed that "science and scientist continue to be governed by fear—fear of public opinion . . . fear of religious intolerance, fear of political pressure, and, above all, fear of bigotry and prejudice." So great was this fear at the time that even some medical textbooks on human physiology lacked entries for *penis* or *vagina*—which is too bad because, when it comes to sex, the human is a fantastically strange animal worth studying in fine detail. While most mammals use sex only for reproduction, humans, for reasons we still don't fully understand, have evolved to use sex for recreation as well as procreation. And that has made our lives much more exciting than those of our ape cousins.

When a female baboon is ovulating, the skin around her vagina swells and turns bright red so male baboons can see it from a distance. In case the males are not looking her way, she also gives off a distinct smell. And if the bright red skin and strong smell don't work, the female will squat in front of the male and present her hindquarters. She knows when the time is right for sex, and she knows how to make it happen.

Such behavior is the norm among mammals. Humans are the strange ones. We're the ones who ovulate with almost no discernible clues. We're the ones who have sex at random times rather than waiting for the time around ovulation (also referred to as *estrus*) when pregnancy is most likely. When a female Barbary macaque is fertile, she'll have sex every seventeen minutes, getting it on at least once with every adult male in her troop. Gibbons go several years at a time without sex while waiting for the female to wean an infant and come into estrus. After a month of abstinence, female baboons will copulate up to one hundred times when they're fertile.

Most animals have sex because they want—or, rather, need—to procreate. Anything else would be a waste of time, and possibly dangerous, because they become vulnerable to attacks by predators when distracted by their mates.

So why do men and women have sex all the time, even when (make that *especially* when) we know fertilization is impossible? Anthropologists have long trumpeted one theory: that the human female has a difficult time raising her offspring alone (and had an even more difficult time in prehistoric days), so she keeps her man around by offering him sex whenever he wants it, even after she reaches an age when she can no longer reproduce. But not everyone buys that argument, and there are a lot more questions that still have scientists scratching their heads. For example, why do humans copulate in private when all other mammals do it in the open? And why do men have bigger penises, in proportion to their bodies, than their ape cousins?

For centuries, the beginning of life was a mystery. Everyone knew that a man had to ejaculate into a woman's body to achieve conception, but beyond that the process involved a lot of guesswork. Most anatomists up to the time of the Renaissance believed people came not from eggs but from seeds (*semen* is Latin for "seed"). Hippocrates believed that conception required two seeds, a male and a female. Aristotle maintained a century later that human life began when the man's seed mixed with the woman's menstrual blood. The debate went on for almost two thousand years. Throughout that time, most people believed that an orgasm was required to generate the heat that a seed or seeds needed to spring to life. The woman had to have an orgasm too, this theory went, given that the conception occurred within her body. It was not until the seventeenth century that the Englishman William Harvey suggested that people come from eggs, and it took yet another two hundred years before scientists figured out that women ovulated monthly.

The science of reproduction might have advanced more swiftly if a few of the researchers involved had been women, but bias was not solely a feature of scientific research. Throughout most of human history, men and women have seldom been treated as equals where sex comes into play. In the Old Testament, when Sarah could not bear children for Abraham, Abraham took a maidservant for a mistress. King Solomon not only had hundreds of wives but had hundreds of concubines, too. In imperial Rome, a woman guilty of adultery was exiled from her home and banned from marrying again. Roman Catholic doctrine declared that sexual intercourse was only for procreation and that thinking or acting otherwise was a sin. In the sixteenth and seventeenth centuries, promiscuous women were burned at the stake. In Victorian England, women were told they were not supposed to enjoy sex, and men were encouraged to visit prostitutes rather than defile their own wives. To discourage promiscuity, birth control and abortion were outlawed in many countries, including the United States, and women were often forced to rely on illegal abortions to control family size. Not until the early twentieth century did anyone dare suggest that sex should be accepted and even embraced as healthy or something to be enjoyed by both men and women.

American attitudes toward sex took a big turn in 1909, when Sigmund Freud gave a series of lectures at the school that would briefly and halfheartedly take in the exiled biologist Gregory Pincus some thirty years later: Clark University in Worcester, Massachusetts.

Born in 1856 in the Austrian town of Freiberg, in what is now the Czech Republic, Freud studied medicine and specialized in nervous and brain disorders. He was influenced by the work of a Viennese colleague, Josef Breuer, who found that he could help deeply troubled patients by getting them to speak openly about the earliest occurrence of their symptoms. Freud theorized that many neuroses were rooted in trauma that had often been forgotten and hidden from consciousness. If patients could be helped to recall their experiences, he suggested, they could rid themselves of their neurotic symptoms.

In 1900, Freud published *The Interpretation of Dreams.* The unconscious mind was a powerful force, he proclaimed, and sexual drive was the most powerful of all determinants of a person's psychology. Sexual urges required gratification, Freud wrote; abstinence was both unnatural and potentially harmful. In Europe, critics complained that Freud was making too much of sexuality, and the good doctor came to be despised. But upon arriving in America he found a welcome and influential audience. "Don't they know we're bringing them the plague?" Freud asked his fellow analyst Carl Gustav Jung as the two men stood on the deck of their ship, staring down at the cheering throngs awaiting their arrival.

Most Americans never bothered to read Freud, but they came to understand, correctly or not, that he had endorsed sex as a desire equal in importance to hunger or thirst. His followers argued that sexual satisfaction was essential to happiness and mental health. Young women in particular, recalled the writer Malcolm Cowley, "were reading Freud and attempting to lose their inhibitions." Freudians did not worship Freud; they worshiped intercourse and orgasms. Among the believers, nothing satisfied desire and made the world a better place more than a mind-blowing, spine-shivering orgasm, or "la petite mort" (the little death), as the French called it, suggesting a mystical quality to sex.

Margaret Sanger took up the cause, and so did Wilhelm Reich, another disciple of Freud. In 1923, Reich told the Vienna Psychoanalytic Society that he believed orgasm was the key to curing neuroses. "Genital stagnation," he warned, would lead not only to emotional problems but also "heart ailments . . . excessive perspiration, hot flashes and chills, trembling, dizziness, diarrhea, and, occasionally, increased salivation." Women and adolescents were particularly vulnerable, he said, because they were expected to remain abstinent (at least until marriage, for women) while men were free to satisfy their sexual appetites. Reich believed that everyone needed orgasms—and lots of them—to discharge their sexual energy and remain healthy. What's more, he said, unless that energy was released, the world would never achieve progressive political or social reform. It would take nothing less than a sexual revolution—a term of Reich's creation—to create a truly free society. Reich was the prophet of the orgasm. He even devised a special box—the Orgone Energy Accumulator—to help harness orgasmic energy, which he believed circulated in the atmosphere and in the human bloodstream. Norman Mailer, Saul Bellow, William Steig, and many other intellectuals later sat in the box (Albert Einstein considered it but politely declined). Eventually the federal government labeled Reich a fraud, but by then it didn't matter. He had already inspired a generation of believers who would become central players in the sexual revolution.

After Reich came Alfred Kinsey. At first glance, Kinsey did not look like a radical. He wore a bow tie and crew cut as he lectured students at Indiana University, and he liked to invite his colleagues to his home to drink tea and listen to classical music from his impressive record collection. He married the first woman he ever dated and took her camping on their honeymoon so he could collect bugs. Sex interested him because it was a part of nature, but work was his real passion. Kinsey was an entomologist who began his academic career studying gall wasps. Only when his students began asking questions about marriage did he begin reading all he could on human sexuality. Appalled by the scarcity of reliable information, Kinsey began his own studies. A radical empiricist, he viewed everything as quantifiable, whether it was orgasms or sex between humans and barn animals. Armed with nothing more than a notebook and a straight face, he set out to measure and categorize the variety of sexual conduct in America. He started by interviewing his students and soon, with a team of researchers, fanned out across the country.

Kinsey discovered he had a great gift for eliciting elaborate and secret information. By 1947, he was ready to publish his results. Among his findings: sex was good for marriage, masturbation did no harm, homosexuality was more widespread than most people assumed, and men and women cheated on their partners more frequently than most people believed. While others weighed in on whether homosexuals or unmarried sex partners were destined to go to hell, Kinsey reported the facts as science: "Mouth-genital contacts of some sort, with the subject as either the active or the passive member in the relationship, occur at some time in the histories of nearly 60 percent of all males." But Kinsey's most important finding was probably this one: Women desired sex, and not just to make babies. They masturbated, they enjoyed orgasms, and they slept around much the same as men did (although, according to Kinsey, they either did so less often or were less willing to admit it). Either way, Kinsey made Americans feel less shame about sex. He assured them their desires—even the kinky ones—were normal. His book, *Sexual Behavior in the Human Male*—which cost $6.50 (about $63 today), had 804 pages, and was published by W. B. Saunders, an older medical publishing company, in 1948—became a surprise bestseller.

Kinsey inspired young men like Hugh Hefner—who used the furniture in his small Chicago apartment as collateral for a bank loan to launch *Playboy* magazine—to think of sex as something healthy and righteous. Hefner would soon see himself as a kind of Paul Revere in silk pajamas, a messenger of truth and freedom. He urged Americans to treat sex as something they were entitled to enjoy selfishly and ostentatiously, like fast cars, good food, and fine spirits.

Thanks to Freud, Reich, Kinsey, Hefner, and others, humans were more unusual creatures than ever by the middle of the twentieth century. They became fascinated by sex, convinced that it was the ultimate source of rapture. Young men began describing the stations of their sexual achievements in competitive terms such as "first base," "second base," and "scoring," or "going all the way." Everything seemed sexually charged. Even the cars of the day looked like phallic rocket ships—except for the Edsel, which had a grille that resembled a chrome vagina. Scandal magazines reported on the sex habits of the stars. Girlie magazines like *Flirt*, *Wink*, and *Titter* offered crude jokes and luscious pinups. Hollywood in the 1940s turned Betty Grable and Esther Williams into objects of sexual worship.

On the surface, the 1950s appeared to be a time of conformity and conservatism, but it was also an age of fear. Russia had the atomic bomb, so families built underground shelters stocked with canned goods and water to last for years and the Department of Defense hid Nike missiles underground all over the country in case of nuclear attack, from Michigan Avenue in Chicago to the Santa Monica Mountains in Malibu. U.S. Senator Joseph McCarthy launched a ruthless campaign to uncover suspected Communist sympathizers, tarring innocent, law-abiding citizens in the process. For women, it was an especially challenging time. They risked being seen as outcasts if they graduated from college without being married, got married and did not immediately have children, or had children but also wanted to work outside the home. To have a child out of wedlock was the greatest of shames.

Even women's clothing was restrictive. "Fifties clothes were like armor," wrote Brett Harvey in the introduction to *The Fifties: A Women's Oral History.* "Our ridiculously starched skirts and hobbling sheaths were a caricature of femininity. Our cinched waists and aggressively pointed breasts advertised our availability at the same time they warned of our impregnability." Nursing and teaching were the only professions easily accessible to women. A woman's role in life was to be married and raise children, and to start at an early age. She was supposed to find satisfaction in serving her husband and her children. If she had desires of her own—be they sexual, professional, or personal—she was expected to hold them in check, to wipe them out the same way she wiped germs from the kitchen counter or stains from the collars of her husband's white dress shirts. To rebel against these restrictions was to invite scorn and humiliation. The unmarried life was seen as empty and joyless, and women living it were to be pitied.

Women in the 1950s tended to marry as soon as they could. The median age of marriage for a woman in 1950 was 20.3. A decade earlier, the median age had been 21.5 (today it is 26.1). Why were young women of the 1950s in such a hurry to get hitched? With the war over and men returning home, single women had few options. They couldn't compete with the men for jobs, and college, while potentially enlightening, only postponed the realization that career options for women were limited. "What's college?" asked an ad for Gimbels department store. "That's where girls who are above cooking and sewing go to meet a man so they can spend their lives cooking and sewing." Another reason to marry: They wanted to have sex, and it was dangerous to do so out of wedlock. Condoms were sold in drug stores, but to get a diaphragm in most states required a doctor's prescription, and most unmarried women were ashamed to ask.

"I knew birth control existed, but I didn't know anything about it," one woman told Harvey for her oral history. "To go out and actually get it [birth control] would mean that I planned to do these things, to have sex. Since I knew it was wrong, I kept thinking I wasn't doing it, or I wasn't going to do it again. Each time was the last time. Birth control would have been cold-blooded."

"I was terribly frightened about getting pregnant," another woman admitted, "but I never did anything about getting birth control. I'm not really sure why. Maybe I kept telling myself we weren't going to do it again."

Soon, of course, the young brides as well as the brazen few who engaged in premarital sex did get pregnant. Not just once but over and over again. As the Baby Boom began and families grew, women raising four, five, or six children began seeking more effective means of contraception. Women who married at nineteen or twenty were done—or wished to be done—with babies by the time they were thirty. Most American women, with the exception of Catholics, accepted the idea of birth control, and most of them wished for a more convenient and effective method.

Fear of pregnancy was an unavoidable part of sex for young women in the 1950s. A woman who was unmarried and pregnant was in terrible trouble. Single motherhood was not an option, at least not among the middle and upper classes. Abortion was illegal and underground abortions could be dangerous or difficult to obtain, especially for those without money. Many women felt trapped—by their bodies, by their career options, by their contraceptive options, by pregnancy, and perhaps most of all by their limited choices.

That's why Margaret Sanger was so interested in meeting with Gregory Pincus. She was seventy-one years old, well past her sexual prime, and had lost some of her brazenness. Instead of fighting for sexual liberation, she employed more pragmatic arguments, touting the importance of population control and family planning.

She had long held that it was not a question of principle but a question of methods. If the right method of birth control were discovered, she believed, the sex—and everything else—would take care of itself.

THREE

Spontaneous Ovulations

THE RABBITS WERE kept in the basement of the Worcester Foundation, along with the rest of the animals, so that their smell wouldn't stick to everyone and everything. Using a small eyedropper, Chang began feeding the animals small amounts of liquid progesterone—between two and five thousandths of a gram.

Chang was tan and slender with thick black hair that he oiled and combed back from his brow. When he smiled, which he often did, a crooked front tooth protruded; otherwise, he was as handsome as a Hollywood leading man, if Hollywood in the 1950s had had Chinese leading men. In China, Chang had won a national competition to earn the right to study abroad. He chose the University of Edinburgh, where he majored in agricultural science and took particular interest in sheep sperm. In part because he spoke English so poorly and in part because it was his nature, Chang came to believe that the key to success was working harder than anyone around. The fact that he was smarter than almost anyone around didn't hurt, either.

Chang spent seemingly endless hours in the laboratory, never complaining. But in truth he did not care much for the progesterone work. Every time an animal was tested it had to be killed and cut open to see if any eggs had been released. It was grisly and inefficient, but Chang refused to delegate the work to an assistant. "I like to feel the experiments through my hands," he once said. "Would you let someone else play tennis or chess for you?"

The initial results, recorded in the spring and summer of 1951, were as he and Pincus had expected. The animals receiving progesterone did not appear to ovulate.

"Victory!" shouted Chang.

Next, Chang tried inserting the hormone in the rabbits' vaginas. That worked, too, although not as well. Larger doses were required, and the progesterone stopped doing its job after about five hours. After the vaginal tests, Chang tried pellets lodged under the rabbits' skin. This time, a single pellet prevented ovulation for months.

Pincus was pleased, but he wasn't finished. Rabbits are not like humans; female rabbits have to copulate to release eggs. So Pincus told Chang to move on to rats, which, like humans, ovulate spontaneously. Rats offered another benefit for research purposes: they're sexually prolific. When a female rat is receptive, she can mate as many as five hundred times with various males in a span of six hours.

Chang caged male and female rats together, two males to every five or six females, and injected some of the females with progesterone. Once again, the experiment worked; there were no pregnant rats. And once again, larger doses had longer lasting effects.

Pincus and Chang ran tests through the night and into the early morning in their first weeks and months experimenting with progesterone, hoping that a solid report to Planned Parenthood might get them more money. Sometimes church groups or Rotary Club members would visit the Foundation to see what kind of work went on. The Foundation was funded in large part by its neighbors, after all, and so Pincus made it a point to welcome tours.

Visitors might find Goody Pincus weighing a female rat's uterus, castrating male rats, or seated behind his desk smoking Viceroys and looking over the budget. He seldom smiled and almost never laughed, but he had an easygoing way that put people at ease. If his visitors ventured into the basement they would see dozens of rabbits and rats, although they probably wouldn't see them having sex because the animals were shy around humans. Pincus enjoyed explaining science to the uninformed. Moreover, he deemed it part of his job. Margaret Sanger wanted a pill, but Pincus was not embarking on this project simply to satisfy a client. He took himself too seriously as a scientist to do straight work for hire. "The modern-day investigator," he once wrote, "cannot be satisfied with the invention of a 'cunning device.'" Tinkering with the reproductive process could be dangerous. A misstep at any point in the process could cause lasting and profound "physiological consequences that are not apparent on the surface." The researcher, he said, must first understand as much of the process as possible, and then he must work to explain that process to others. He mocked as naïve the "ivory tower conception of research" that says a scientist should do his research, publish his results, and wash his hands of the matter. The modern world required a different, more activist brand of science, he said. It would not be enough merely to create a more effective contraceptive. If such a thing were to work, the scientist leading the research would have to make sure doctors, nurses, clinicians, and patients understood the how and why of it. He would have to be an evangelist. He would have to see that the contraceptive was properly used, just as the physicists who worked on the atomic bomb had done. They didn't hand off their bomb and move on; they formed safety committees and promoted dialogue about the weapon's future use. Pincus couldn't understand why physiological researchers weren't more engaged with the world in which they lived.

※

In the 1930s and even the 1940s, contraception was controversial and hormone research was in its primitive stages. But by the time Pincus and Chang came along, the world was changing. Many politicians, journalists, intellectuals, and social activists viewed population growth as a threat to economic development and world peace. Between 1920 and 1950, poor countries had been growing much more rapidly than prosperous ones. There was a growing sense among activists and intellectuals—a sense often informed by racism, arrogance, and politics as well as genuine concern—that high birth rates in poor countries would devastate the world. Poverty and starvation would spread; the diseased and deficient would multiply; and overpopulated nations, in desperation, might tip to communism. In 1927, a Rockefeller Foundation–funded study of contraception sought "some simple measure which will be available for the wife of the slum-dweller, the peasant, or the coolie, though dull of mind." In language that was widely accepted at the time, some argued that governments should subsidize the sterilization of the feeble-minded as well as people with communicable diseases.

In 1932, the novelist Evelyn Waugh warned in his book *Black Mischief* that finding solutions to population growth would not be as simple as crusaders like Sanger hoped. The novel's hero, an English playboy living on a tropical island, designed a poster meant to discourage couples from producing big families. The poster displayed two scenes: in one, a family with eleven children manifested signs of disease and malnutrition; in the other, a husband and wife with one child lived in affluence. Between the two pictures was the image of a contraceptive device and the legend "Which home do you choose?" The islanders in Waugh's book chose the larger family and concluded that the device in the middle—"the Emperor's juju"—was responsible for the unfortunate condition of the couple that had only one child.

Changing such attitudes would never be easy. Sanger supported economic development and education. At the same time, for all her tireless efforts as a champion of women, she could be shockingly insensitive, too. She agreed with the eugenicists who said that women not qualified to be mothers ought to be sterilized. But sterilization, education, and economic development were not enough. She sought a solution that would do it all—reduce population size, restrict reproduction among unfit parents, and make sex more fun, and she had come to believe that only a truly scientific contraceptive would do. A scientific solution would give her the legitimacy she needed to make a broad and lasting impact.

If Sanger had approached Pincus with the idea of developing a pill solely to allow women more pleasure in sex, it's unlikely that he or any other male scientist would have risked his reputation on it. But now he had a chance to create a simple solution to many of the world's most daunting problems. These were Sanger's longtime concerns, not his, but he could see the potential. When he began, he was interested primarily in the science, but he quickly understood the social change a birth-control pill could effect. "Our globe is facing a threat that could be far more serious than the atomic bomb," he told one journalist. Birth control struck him as an issue big enough to bring him the fame and respect he believed he was due.

※

The Worcester Foundation, with about twenty scientists, operated on an annual budget of $300,000. Residents contributed about $63,000 of that amount. Forty miles west of Boston, Worcester had a population of 208,000. It was a booming factory town in which about six hundred and fifty companies employed nearly fifty thousand men and women in the manufacture of steel, wire, machine tools, grinding wheels, coiled springs, carpets and rugs, corsets, shoes, envelopes, leather goods, woolens, skates, automobile parts, firearms, boilers, sprinkler systems, wrenches, crankshafts, wool-spinning machines, and electric clocks. The city had more than thirty hotels, ten theaters, two daily newspapers, and a prestigious art museum containing important works by Renoir, Monet, and Gauguin. Worcester residents were proud to live in one of the biggest manufacturing cities in the country not located on a waterway. They were also proud, thanks to Pincus and Hoagland, to have their own scientific foundation, which they supported in much the same way they supported the local Boys Club. One year, supporters of the Foundation sponsored a barbershop quartet concert at Mechanics Hall that raised five hundred dollars. Pincus and other Foundation leaders gave dozens of lectures each year to community groups and social clubs. Local businesses like the Wear-Well Trouser Co. and the Worcester Baking Company pitched in with donations. But as the Foundation grew and as support for scientific research expanded in the years after the war, community support was eclipsed by government grants and drug company contracts.

Pincus and Hoagland were fortunate to launch the Worcester Foundation at a time of enormous growth in the pharmaceutical industry. The catalyst was the discovery in the 1930s of the first commercially available antibacterial drugs, known as sulfas, followed by the introduction of penicillin as a drug in the early 1940s. By the late 1940s and early 1950s, drug makers like G. D. Searle & Co. were no longer content to manufacture familiar products; they were competing fiercely to discover and market new ones. In the late 1940s, Searle, a small pharmaceutical company based in Skokie, Illinois, and other drug companies were looking for a way to synthesize cortisone, which had recently been demonstrated to relieve arthritis pain. Pincus persuaded the drug company that he could synthesize cortisone by pumping serum through the adrenal gland of sows—a method referred to as perfusion—and spent half a million dollars of Searle's money trying to prove it. But before Searle could make use of Pincus's new technology, which was effective up to a point, researchers at the Upjohn Company in Kalamazoo, Michigan, found a simpler and cheaper way to do the job.

In the fall of 1951, hoping to repair the relationship with Searle and secure their help on Margaret Sanger's progesterone project, Pincus went to Skokie to meet with Albert L. Raymond, the drug company's director of research. Raymond, a small, studious man with a thin, red mustache, told Pincus that his most important benefactor was losing faith. Though Sanger and Planned Parenthood had invited Pincus to work on a new contraceptive, the money for that project was paltry and might dry up at any moment. He needed Searle. Yet his meeting with Raymond did not go well. When it ended, he was so rattled that he grabbed several sheets of hotel stationery and wrote Raymond a frantic letter.

"Since sleep escapes me," he began, "I will try to set down what I think is a fair summary of what you said to-night as we were driving around. You said: 'You haven't given us a thing to justify the half-million that we invested in you . . . and the responsibility for this failure is yours. . . . To date your record as a contributor to the commerce of the Searle Company is a lamentable failure, replete with false leads, poor judgments, and assurances from you that were false. Yet you have the nerve to ask for more.'" After summarizing Raymond's comments, Pincus framed his response, one that was both professionally and personally close to groveling, revealing a kind of doubt and desperation he almost never permitted anyone to see. "I feel that the moral is plain," he wrote. "There should be, from a business point of view, no need for further support of a person with such a record."

Pincus had not merely tried and failed, he had tried hard and failed badly. The loss of Searle's support would be a huge blow to the Worcester Foundation. Already he was having a difficult time paying workers what they were worth. Only loyalty and love of the work kept his top scientists from taking better jobs. Now it was possible that he would have to dismiss workers or encourage them to take jobs elsewhere. For Pincus personally, the failure was profound, leaving him to wonder if he would ever achieve the greatness of which he believed he was capable and if the Worcester Foundation's days were numbered.

"I want you to know," he wrote to Raymond, "that I have indeed been embarrassed at the failure to see a paying result. I have done what I could, but it is obviously in your view no good. My attempts have led me into a situation which is rather difficult. . . . [N]ow at a time when I am just about at the peak of productive activity I see my wife buying $6.95 dresses the way she did when we were first married . . . and if I were to die I would leave my family not too well provided for."

The letter is neither an apology nor a plea for forgiveness. It reads, instead, like the work of a passionate scientist, one who has analyzed the data carefully in an attempt to explain his own failure and its consequences. Given his uncertain status with Searle, it was no wonder Pincus would be reluctant to turn down Margaret Sanger's "ludicrous" offer of $2,000 to fund birth-control research. He was in no position to turn down anything.

FOUR

A Go-to-Hell Look

MARGARET SANGER HAD always been good at bending men to her will, exciting them with her energy and ideas, and then leading them into battle—or into bed.

With nothing more than a few words of encouragement and a vague offer of money presented at their first meeting, she had sent Gregory Pincus speeding back to his lab in Worcester, Massachusetts, ready to go to work. Yet despite her seemingly infinite powers of persuasion, there was no telling if she'd be able to get Pincus or anyone else to produce the birth-control pill that had so long been her wish.

"I've got herbs from Fiji which are said to be used to prevent Conception," she wrote to a friend and supporter in 1939. "I'm hoping this may prove to be the 'magic pill' I've been hoping for since 1912 when women used to say 'Do tell me the secret.'" And the wish was not only Sanger's; she was merely the torch carrier for millions of women around the world. She received letters such as this one:

Englishtown, N.J.
January 5, 1925

Dear Mrs. Sanger

I received your pamplet [sic] on family limitation. . . . I am 30 years old have been married 14 years and have 11 children the oldest 13 and the youngest one year. I have kidney and heart disease, and every one of my children is defictived [sic] *and we are very poor. Now Mrs. Sanger can you please help me. I have miss* [sic] *a few weeks and don't know how to bring myself around. I am so worred* [sic] *and I have cryed my self* [sic] *sick and I don't come around I know I will go like my poor sister she went insane and died. My Doctor said I will surely go insane if I keep this up but I can't help it and the doctor won't do anything for me. Oh Mrs. Sanger if I could tell you all the terrible things that I have been through with my babys* [sic] *and children you would know why I would rather die than have another one. Please no one will ever know and I will be so happy and I will do anything in this world for you and your good work. Please please just this time. Doctors are men and have not had a baby so they have no pitty* [sic] *for a poor sick Mother. You are a Mother and you know so please pitty me and help me. Please Please.*

Sincerely your

[J. M.]

After three decades of searching for something better to offer the desperate women who wrote to her, Sanger had nothing. In 1950 her magic pill remained a dream, blurry and out of reach. All her life Sanger had led a fight few others would dare. But now she was old and in poor health. She'd survived a heart attack in 1949, and now she looked better suited for the deck of a Caribbean cruise ship than a picket line. At times, even some of her supporters questioned whether she was losing her edge.

She would go on, but the question was for how long.

※

Sanger herself had come from a family of eleven children. She was born Margaret Higgins in 1879 in the factory town of Corning, New York. Friends and family called her Maggie. Her mother, a frail and submissive woman, died of tuberculosis at age fifty. Her father, Michael Higgins, was a Civil War veteran and a stonecutter who chiseled the angels for the tombs in the town's cemetery. He worked in a big barn of a studio where men would gather at the end of the day to talk. Michael Higgins was such a good talker that his daughter—his sixth child—liked to imagine that he could chisel away people's outer edges and open them up, revealing the true angels within. His friends loved him and trusted him, but Maggie's own feelings were complicated, for she always believed that her mother had fallen victim to her father's sexual appetite. Eleven children had been too many to bear.

Maggie, too, feared her father's passion. In an account of her life published in 1931, she described lying in her mother's bed as a little girl, boiling with a case of typhoid fever, and feeling a man lay down beside her:

> *It was Father. I was terrified. I wanted to scream out to Mother to beg her to come and take him away. I could not move, I dared not move, fearing he might move toward me. I lived through agonies of fear in a few minutes.*

Sanger wrestled with feelings about her father that she couldn't articulate. She adored him for his warmth, bravery, and independence. She thanked him for teaching her to be bold and brave and to challenge orthodoxy and narrow-mindedness wherever she found it. Michael Higgins was a Catholic apostate. Once, Higgins invited Robert Ingersoll—a freethinking secularist known to his admirers as the "Great Agnostic" and to his critics as "Robert Injuresoul"—to speak in the heavily Catholic town of Corning. Ingersoll dined with the Higgins family. After eating, Ingersoll and the Higginses walked under gray skies to Town Hall, where Michael had made arrangements for the speech to be held, only to find an angry swarm of people and a constable barring the door. Twelve-year-old Margaret watched her father turn and speak to the crowd, saying those who wished to hear Ingersoll speak should walk with him to the edge of town. Higgins and Ingersoll then led a three-mile procession, with Margaret and a few neighbors trailing behind. It was late in the afternoon when they climbed a hill and stood beneath a lone oak, where Ingersoll finally spoke.

After that, Sanger recalled, the Higgins children "were known as children of the Devil . . . the juvenile stamp of disapproval had been set upon us." Her father would no longer be hired to carve angels for the Catholic cemetery, and Maggie would never forget that the Catholic Church had unfairly branded her a sinner.

But she would take away another lesson, as well: "I have always known that when they said they could stop us from speaking, they were wrong." Maggie broke free. With financial support from her older sisters, Mary and Nan, who worked as a maid and a governess, respectively, at wealthy homes in Corning, Maggie left home and enrolled at Claverack College, a boarding school in New York's Hudson Valley. At Claverack, she worked in the school kitchen to earn her room and board but also began speaking out on suffrage and women's emancipation, which were radical issues at the time. When her mother died in 1899, worn out by consumption at the age of fifty, Maggie might have been expected to return home and inherit a slew of household chores. She might have spent the rest of her life in a role of forced domesticity, shuffling along from subservient daughter to subservient wife to subservient mother until her fighting spirit was gone. But again she escaped, thanks in part to financial support from her sisters. Maggie Higgins enrolled in nursing school at the White Plains Hospital in Westchester County, New York. "I wanted a world of action," she wrote. "I longed for romance, dancing, wooing, experience." Marriage was not on that list, but sex certainly was. She began to think of sex as something more than recreational—a path to self-improvement, a source of health and happiness, and perhaps even liberation.

Hers was a philosophy born of a strong libido and a strong mind. Maggie Higgins read *Love's Coming of Age*, by Edward Carpenter, who equated the power of sex with religious consecration. She discussed with friends the radical new ideas of Sigmund Freud, who believed that sex was the most important determinant of self. The removal of sexual restraints, said the great Austrian founder of psychoanalysis, would liberate the soul and allow women to more fully experience life and all its joys. Maggie agreed. There was only one hitch, as far as she was concerned: Freud never addressed the fact that sex could result in pregnancy as well as liberation, and that the two were not entirely compatible.

Though she considered marriage "akin to suicide," at the age of twenty-two Maggie met a man at a hospital dance, a handsome young painter and architect named William Sanger, the son of German Jewish immigrants. She described their early romance in a manner that may have suggested her ambivalence toward courtship and marriage: "On one of our rambles he idly pulled at some vines on a stone wall, and then, with his hands, tilted my face for a kiss. The next morning, to my mortification, four telltale finger marks were outlined on my cheek by poison ivy blisters. . . . I was sick for two months." She recovered; they fell in love, married, and built a home in Hastings-on-Hudson in Westchester County, New York. Soon came three children, two boys and a girl. Sanger, unsurprisingly, found no bliss in suburbia or in marriage. In 1912, the family moved to New York City, and Margaret began spending time in a place better suited to her rebel soul: Greenwich Village.

The Village was packed with radicals and rogues. Longshoremen occupied barstools next to poets. The patron of the arts and activist Mabel Dodge held salons with guests she described as "Socialists, Trade Unionists, Anarchists, Suffragists, Poets, Relations, Lawyers, Murderers . . . Newspapermen, Artists, Modern Artists." It was in the Village that Sanger met the celebrated Socialist leader Eugene

Debs and the feminist agitator Emma Goldman, who became a mentor to her. It was there that she heard Big Bill Haywood discussing the Industrial Workers of the World and Walter Lippmann sharing his thoughts on Freud. Some radicals, influenced in large part by Freud, urged women to fight for more than the vote. They wanted a complete change in values and attitudes toward the role of women in society. They wanted to make sexual freedom a part of broader social reform. They wanted motherhood to be voluntary. Sanger went further: she thought sex should be at the *center* of any reform.

"I love being swayed by emotions," she wrote in her journal in 1914, "just like a tree is rocked to and fro by various breezes—but stands firmly by its roots." Sanger wanted women to have more autonomy in the bedroom and in society. She wanted them to see sex as something vital to their senses of identity and self-expression. She became the most outspoken advocate of sexual pleasure the country had ever seen.

"It was as if she had been more or less arbitrarily chosen by the powers that be to voice a new gospel of not only sex knowledge in regard to conception, but sex knowledge about copulation and its intrinsic importance," Mabel Dodge said. "She was the first person I ever knew who was openly an ardent propagandist for the joys of the flesh."

While undergoing this education in radical thought, Sanger worked for Lillian Wald's Visiting Nurse Service, a group of nurses sent out by the Henry Street Settlement House to care for poverty-stricken women and, often, help the women through childbirth. She found the conditions "almost beyond belief." She wrote: "I seemed to be breathing a different air, to be in another world and country." At the time, more than six hundred thousand people crowded below Fourteenth Street east of Broadway. There was New Israel, Little Italy, Hell's Kitchen, and the "Bloody Sixth" ward, all of them teeming with poor immigrants. In 1910, in one fairly typical tenement at 94

Orchard Street, sixty-six people occupied eight apartments of about 460 square feet each. Between 1890 and 1910, Manhattan's population had increased 62 percent, to about 2.3 million from 1.4 million. Russian Jews and Italians led this massive wave of immigration. Sanger was astonished by the poverty and misery: children sick, dirty, and underfed; tuberculosis rampant; and women seemingly unaware of how their own bodies worked and the risks of repeated pregnancies and venereal disease.

"Poor pale faced wretched wives," she wrote to a friend. "Men beat them they cringe before their blows but pick up the baby—dirty, & ill-kempt & return to serve him." She watched women die because their bodies could not hold up against the strain of producing so many babies in such poor conditions, or because they used primitive birth-control devices that caused infection, or because butchers posing as abortionists botched their jobs.

In the 1920s, the state health department of New York distributed circulars warning women that pregnancies occurring too close together were dangerous, predisposing mothers to tuberculosis. But the same department barred women from receiving information about how to prevent pregnancy. Doctors estimated that one-third of all pregnancies in the United States at the time ended in abortion. Sanger saw poor women resorting to "turpentine water, rolling down stairs, and . . . inserting slippery-elm sticks, or knitting needles or shoe hooks in the uterus" to end their pregnancies, and that helped her find a focus for her fury.

She told the story of one woman's death that hit her especially hard. The woman's name was Sadie Sachs. Her doctor had warned her that another pregnancy might be fatal, but the doctor's only advice to prevent it was for Sachs to sleep on the roof, away from her husband. She got pregnant again and died after an attempted abortion. Sanger said it was this death more than any other that compelled her to resolve that women should have the right to contraception. "I would strike out," she wrote. "I would scream from the housetops. I would tell the world what was going on. . . . I would be heard. No matter what it should cost. I would be heard." In 1913, she wrote a twelve-part series of articles about sex and reproduction for *The Call*, a radical newspaper. The articles carried the most straightforward headline she could conjure: "WHAT EVERY GIRL SHOULD KNOW."

While Sanger battled on behalf of women struggling to raise more children than they could handle, she was strangely indifferent toward her own children. She had become pregnant about six months after her wedding. Suffering at the time from tuberculosis, she found that pregnancy aggravated her symptoms, and she was forced to spend most of the pregnancy at the well-known Trudeau Sanitarium in Saranac Lake, New York. After the birth of her first son, Stuart, in 1903, Sanger lapsed into severe depression. It would be five years before her second son, Grant, was born. Twelve months later came a daughter named Margaret. Afraid her children would contract tuberculosis from her, she hired nurses and nannies to supervise their care. She was easily annoyed by the petty rivalries of her children, her biographer Ellen Chesler wrote, and kept them at a "predictable distance." When her son Grant was ten years old and away at boarding school, he wrote to ask if he should return home for Thanksgiving, as most of his classmates were preparing to do. Sanger wrote back that, yes, he should come home; the maid would be there to cook him dinner.

Sanger consciously rejected the idea of full-time motherhood, making up her mind that her work was too important to compromise, and she never expressed regret over the decision. By the time she became a mother and a full-time activist and began to test the bounds of her own independence, American women had acquired a measure of power over sex and reproduction within marriage, even if it was hardly evident in the tenements in which Sanger spent so much of her time. In fact, the shift in power was so subtle that many women had not yet noticed it. During the nineteenth century, 90 percent of women were married and 95 percent were not employed outside the household. Over the course of the century, though, there was one big change for women: their fertility rate dropped by 50 percent (the average number of children born to a white woman fell from 7.04 in 1800 to 3.56 in 1900).

Much of history tells us that women of the Victorian era had little control over their lives, but it wasn't so. Though they had no reliable contraceptives, women were nevertheless making private efforts to reduce the sizes of their families and liberate themselves, at least a little, from the demands of domestic labor. There is, after all, a big difference between raising seven children and raising three or four. How did women do it? Simply: by saying "no" more often to their husbands, or by asking their men to withdraw prior to ejaculation. Men still dominated most of society, but women were wresting away control of the home and, increasingly, of sex. As women asserted more power in the bedroom, they extended their influence outside of their homes as well. They became more active in their churches and began organizing to fight for political and social change, beginning with the right to vote and the drive to prohibit the sale of alcohol. One big factor in the push for the ban of booze: women believed their husbands would be less abusive and less likely to force them to have sex if more men quit drinking.

Abortion rates rose dramatically in the nineteenth century. Women experimented with an array of contraceptive devices, but most of them did more harm than good. Dr. Clelia Duel Mosher conducted one of the few studies of contraception in this era, beginning around 1892, when she was a biology student at the University of Wisconsin, and running until 1920. Mosher chronicled the sex lives of forty-five women and found that twenty-eight of them used contraception. Most of the women in Mosher's study were affluent, with money and connections to visit doctors or buy what they needed on the black market. Among her respondents, the most popular form of contraception was Lysol, the antiseptic soap, whose formula in those days contained cresol, a phenol compound that often caused inflammation and burning. The second most popular choice was the condom. One of the women said her doctor, warning that she could not withstand another pregnancy, had prescribed a "woman's shield," a device that was supposed to form a seal around the cervix and prevent semen from getting through. Unfortunately, caps such as these seldom fit properly. If a cap were too large, a woman might suffer cramping, ulceration, or infection; if it were too small, pregnancy was likely. Some early studies showed failure rates as high as 24 percent for these devices. Another one of Mosher's respondents had received an intrauterine device, or IUD, which was one of the newer forms of contraception and one of the most painful and medically dangerous. IUDs in the 1920s were essentially rings made of silkworm gut shaped by silver wire. The rings were bulky and, in the days before antibiotics, sometimes led to fatal infections of the uterus, ovaries, or fallopian tubes. Another option, the douche, had an absurdly high failure rate—as high as 90 percent—and caused frequent infections. Some women used two, three, or even four methods simultaneously, and still they were not safe. The rhythm method—whereby women confined their sexual activity to the time of the month when they were likely to be infertile—was not yet a factor at the time of Mosher's study because scientists did not fully understand how and when women ovulated. As late as 1930, the *Journal of the American Medical Association* had this advice for doctors: "We do not know of any method of preventing conception that is absolutely dependable except total abstinence."

Impoverished women often had no option other than abortions, the tragic results of which Sanger had witnessed. The first known description of an abortion appeared around 1500 BC in a medical text describing how a plant-fiber tampon coated with honey and crushed dates was used to end a pregnancy. Women swallowed lye and gunpowder, placed leeches inside their bodies, poked themselves with knitting needles, threw themselves down stairs, hammered their abdomens with brickbats, and swallowed poisons. They were willing, in short, to risk serious side effects, arrest, and death rather than remain pregnant.

Sanger implored women to fight. She wanted them to control their bodies and decide for themselves whether and when to have children. She wanted them to enjoy sex in or out of marriage; to think about their feelings; to be the rulers of their own hearts, minds, and bodies. Sanger believed in free love, the magic of which she heard described in so many Greenwich Village salons. She may also have believed, as the prevailing myths of her day suggested, that tuberculosis was a disease of both body and soul, ravaging her physically but also infusing her with a stronger capacity for passion than others. Sanger, even in illness and old age, never ceased to see herself as an attractive woman worthy of love and eager to express her sensuality. When her husband complained about her sexual affairs, she encouraged him to have some of his own. Better balance might help their marriage, she explained. But Bill Sanger was not cut out for casual sex. "You are a world Lover," he wrote his wife. "I am not—I am a single lover. I love too deep & not broad enough."

The marriage was shattering, but once again Sanger showed few signs of regret. She wanted more women to have the same confidence she had, the same willingness to defy convention and face the world, as she put it, with a "go-to-hell look in the eyes; to have an idea; to speak and act in defiance of convention." Such boldness was too much to ask of most women raising children, especially women raising children in poverty. But it was in this spirit, with the support of radical feminists, socialists, and sexual free thinkers, that Sanger built her grassroots movement. Her revolution was different from those that had come before. Suffragists had been fighting for something women could do independent of or even in opposition to men: vote. But Sanger wanted women to fight for sex, and sex usually involved men.

This would be a more delicate business. Men would not submit to being strong-armed. They would have to be cajoled, engaged, perhaps even wooed, and not everyone would succumb to Sanger's potent charms.

FIVE

Lover and Fighter

IN 1914, SOON after publication of the first edition of Sanger's newspaper, *The Woman Rebel*, federal Post Office inspectors issued a warrant for her arrest, charging her with four counts of violating U.S. obscenity laws. If convicted, she would face up to forty years in prison. Sanger, now thirty-four years old and a mother of three, chose not to appear in court. Instead, she jumped bail, left her family, and moved to Europe, where she fell in love with Henry Havelock Ellis, one of the world's preeminent sexual psychologists.

Ellis was fifty-five, long and lean with a flowing white beard and thick head of white hair framing a strong, handsome face. He had a tendency to sit in silence when he had nothing serious to say, but Sanger loosened him up quickly. She was a sexy slip of a woman, a redheaded fireball of lust and curiosity, and in Europe she was freer than ever to explore her passions. "To see her, one is astounded at her youth, at her prettiness, her gentleness, her mild, soft voice," wrote Robert Hale of *The New Yorker* in 1925. "One is reminded of Boticelli's Judith—a gentle spring-like maid who treads the hills as if she danced—but who is attended by a maid upon whose shoulders is the severed head of Holofernes." It was the flame burning within Sanger's "fragile container," Hale wrote, that made her so attractive.

Ellis swooned over "her devotion to an ideal, her fire, her vitality and beauty." He would later write that he "had never been so quickly or completely drawn to a woman in the whole of his life."

Ellis had made it his mission to solve the mysteries of sex, collecting histories from men and women in an effort to prove that physical intimacy was natural and varied. He attacked Victorian notions and ridiculed American prudery. Masturbation was good, women were as desirous of sex as men, and marriage needn't have anything to do with one's decision to engage in physical intimacy. Going further, he argued that sex was "the chief and central function of life . . . ever wonderful, ever lovely."

Ellis introduced Sanger to more intellectuals, including the science-fiction writer H. G. Wells, who became another of her lovers and would go on to write novels based on their romance. "It is wonderful enough that we should take food and drink and turn them into imagination, invention and creative energy," Wells wrote in *The Secret Places of the Heart*, a novel that reads like a love letter to Sanger, "it is still more wonderful that we should take an animal urging and turn it into a light to discover beauty and an impulse towards the utmost achievements of which we are capable." In addition to Ellis and Wells, Sanger also met the playwright George Bernard Shaw and the philosopher Bertrand Russell and had an affair with the Spanish anarchist Lorenzo Portet, who was living in exile and teaching at the University of Liverpool.

Ellis, though, was her chief mentor. He helped Sanger focus her anger and take a more strategic approach to her crusades. Contraception was a fine cause, he said, but Sanger would have better luck if she viewed it in simple terms with a single goal rather than as part of a broad assault on capitalism, marriage, and organized religion. Sanger at times had seemed focused primarily on making noise and upsetting her enemies. Her mission lacked clarity. Ellis educated her on the science of contraception and the economic peril of population growth.

He encouraged her to read about eugenics, including accounts of scientific breeding practiced by the American Oneida Community, and showed her a book by George Drysdale called *The Elements of Social Science*, which argued that only contraception had the potential to increase the amount of love in the world. Drysdale may have been the first modern thinker to state that science had the power to make the world a sexier place, and Sanger became an eager disciple.

Sanger already had the drive and determination to change the way men and women lived, but now, for the first time, she had a plan. At first she had seen contraception primarily as a way to help women control the size of their families. Now she was beginning to believe that if sex were something disconnected from childbirth, women might be liberated in ways they'd never imagined. Marriage would change. Human dynamics would change. The meaning of family would change. Career and educational opportunities for women would change. With that, she had found her mission and distilled it to its essence.

※

While Sanger was in Europe, her husband was arrested by Anthony Comstock, a special agent for the postal service, for distributing pamphlets about birth control.

In the years immediately following the Civil War, Comstock had made it his mission to fight smut in America, almost single-handedly creating a strict set of anti-obscenity laws. As a teenager, Comstock had masturbated so obsessively that he thought it might drive him to suicide. When he got a little older, he concluded that it wasn't his fault; it was filthy books and postcards that nearly drove him to degeneracy. With the backing of influential businessmen, Comstock was appointed special agent for the Young Men's Christian Association (YMCA) Committee for the Suppression of Vice. While leading raids and snatching up sex devices, pornographic pictures, and contraceptives, Comstock became famous for guarding America from pornography and disease. In 1873, he persuaded Congress to pass a bill banning the use of mail for transporting "any obscene, lewd, lascivious or filthy book, pamphlet, picture, paper, letter, writing, print or other publication of an indecent character." After that, every state in the union enacted its own anti-obscenity laws, many of which made it illegal to sell or disseminate information on contraception. To help enforce the federal law, Comstock was appointed a special obscenity agent of the United States Post Office. Soon after, he was authorized to carry a gun.

Comstock was bald, bull-necked, and barrel-chested, with thick muttonchop sideburns running along the sides of his broad face. He called himself a "weeder in God's garden," but he looked a lot more like a dangerous hunter, and he never had trouble finding prey. Comstock and his agents began arresting publishers, bookstore owners, and photographers—often after setting them up by posing as buyers of their wares. Comstock was relentless. He went on to make hundreds of arrests and seize 200,000 obscene pictures and 64,000 contraceptive devices and instruments of sexual pleasure. At least fifteen women accused of immorality committed suicide rather than face the charges.

The Comstock Law defined immorality so broadly that it could have included anything, so it was no surprise that it was deemed to ban not only the sale of contraception but also the transmission of information regarding contraception. The only real surprise is that the law would continue to influence policy and keep women from getting birth control for as long as it did.

※

Soon after her arrival in England, Sanger wrote to her husband saying that she considered their twelve-year marriage over. She asked for a divorce. But Bill did not feel the same way. "You are all the world to me," he insisted in one letter.

If she showed little remorse toward leaving her husband, her feelings about her children were different. She expressed worry about them in her diary. Seven-year-old Grant and five-year-old Peggy were staying temporarily in Greenwich Village under the care of Caroline Pratt, a progressive educator, and her companion, Helen Marot, a labor organizer. Peggy was confused and upset by her mother's departure and cried every time her father had to leave after a visit to Pratt and Marot's home, Bill wrote in letters to Margaret. Eventually, Bill brought Peggy home and asked one of the girl's aunts to move in and do the cooking. Grant wrote letters to his mother, stoically telling her not to worry about him, while Stuart, alone at boarding school, asked his mother to send a photograph he could look at in her absence. With the outbreak of war in Europe, the mail had become unreliable. "How lonely it all is," Sanger wrote. "Could any prison be more isolated . . . than wandering around the world separated from the little ones you love . . . ?" But in other letters and diaries she made clear that she cherished this time free of familial responsibility to "reflect, meditate and dream."

In the fall of 1915 Sanger's husband was convicted on obscenity charges, with the judge saying he had not only violated "the laws of man, but the law of God as well, in your scheme to prevent motherhood." Offered a $150 fine or a thirty-day prison sentence, William Sanger chose prison. Only then did Margaret finally agree to return home. Soon after her arrival in New York, her five-year-old daughter Peggy developed pneumonia and died at Mount Sinai Hospital, cradled in her mother's arms. The Sanger family would never be the same. Grant blamed his mother, saying Peggy would never have become so sick if their mother had been present. Bill Sanger sculpted a plaster cast of his daughter that he would keep in broken pieces for years. Margaret suffered sleepless nights, and when sleep did come she was haunted by nightmares. In one dream, she heard roofs crashing around her and began worrying about her daughter, only to realize she'd been neglecting the girl and didn't know where she had gone. For years after she would be tormented by dreams of babies, and sleep would remain a problem for the rest of her life.

Still, this tragedy did not compel her to pay more attention to the care of her surviving children. Instead, Sanger went back to work, determined to achieve something important and ready to wage war if necessary. She began by posing for publicity photos, wearing a wide Quaker collar with her hair pinned neatly and her young son by her side, looking like a respectable young mother. It was about this time she began using the term "birth control" instead of contraception, a brilliant piece of marketing strategy. Sanger wanted to separate sex from reproduction, but there was more to her movement than that. At first, she considered referring to her cause as "voluntary parenthood" or "voluntary motherhood," but those terms didn't quite ring true. "Then we got a little nearer," she said, "when 'family control' and 'race control' and 'birth rate control' were suggested. Finally it came to me out of the blue—'Birth Control!'"

There was no sexual connotation involved, no declaration of independence, no threat. These were not fighting words. These were words, like her Quaker collar, designed to make people more comfortable. No one could object to "birth," of course. Without birth there could be no life. But for Sanger, the key word was "control." If women truly got to control when and how often they gave birth, if they got to control their own bodies, they would hold a kind of power never before imagined. Without control, women were destined to be wives and mothers and nothing more. But the word "control" also sent the message to Sanger's supporters in the eugenics movement that she had their mission in mind, too—that this was not only about personal choice but also about determining who should and should not be giving birth.

In 1916, Sanger opened her first birth control clinic at 46 Amboy

Street in the Brownsville section of Brooklyn, where she, her sister Ethel, and a team of nurses distributed condoms and Mizpah pessaries (flexible rubber caps that were commonly sold in drug stores as a "womb support" but in reality functioned similarly to a diaphragm or cervical cap). The pamphlets advertising the clinic, printed in English, Yiddish, and Italian, read:

MOTHERS!
Can you afford to have a large family?
Do you want any more children?
If not, why do you have them?
DO NOT KILL, DO NOT TAKE LIFE, BUT PREVENT
Safe, Harmless Information can be obtained at
46 AMBOY STREET.

Given that the clinic operated in direct violation of New York State law, no one was surprised when police raided the place ten days after it opened, confiscating contraceptive devices and arresting Sanger. She served thirty days in the Queens County Penitentiary, charged with the illegal distribution of birth-control products. Arguing that she was a political prisoner, not a real criminal, she refused to be fingerprinted. Before her release, a pair of prison guards named Murray and Foley tried to force her to submit to fingerprinting, but Sanger, who must have been outweighed by a three to one margin, fought them off. The *New York Tribune* reported that she emerged from jail to greet her supporters with "wrists reddened as though they had been rubbed vigorously." Another newspaper slyly added, "Nobody outside the Corrections Department knows what scars Murray and Foley are nursing."

The timing of all her publicity was excellent. Even people not concerned with women's rights or free sex were beginning to suggest that birth control might work for the greater good of society. In

1798, an obscure English clergyman named T. R. Malthus published "An Essay on the Principle of Population," arguing that most of the world's suffering was caused by mankind's "constant tendency . . . to increase beyond the means of subsistence." Charles Darwin believed that humans were becoming more fertile over time, and Margaret Sanger's lover, Havelock Ellis, extrapolated that man's sex drive had accelerated now that he was less preoccupied with the physical struggle for survival. The rise of cities and the Industrial Revolution, not to mention the Catholic Church's strictures, contributed to growing family sizes. And while birth rates grew, death rates fell thanks to advances in science and medicine. For the first two to five million years of human history, the world's population never topped ten million, and population growth was scarcely above zero. Birth and death rates were almost equal. But when humans began to farm and raise animals, they began living longer. By 1658, the world population had reached five hundred million. By the year 1800, it had topped one billion. And by 1900, the number was approaching two billion.

In the United States, where land was abundant, population control was not a pressing economic problem. But it was becoming a pressing social problem. Children from large families were more likely to be forced to leave school and go to work. Families with a dozen or more children often clustered into three or four rooms until, inevitably, some of the children died or became delinquents and moved out to live on the streets, where they often turned to crime. Thousands of men and women with syphilis went on reproducing infected children. One such inflicted woman wrote to Sanger:

> *I am today the mother of six living children, and have had two miscarriages. My oldest son is now twelve years old and he has been helpless from his birth. The rest of my children are very pale and I have to take them to the doctor quite often. One of my*

> *daughters has her left eye blind. I have tried to keep myself away from my husband since my last baby was born, but it causes quarrels, and once he left me saying I wasn't doing my duty as a wife.*

By the 1920s, Sanger had many new allies. More than 4.7 million American men, many of them from working-class or immigrant families, served in the military during World War I, where they learned about venereal disease and condoms from their fellow soldiers and from prostitutes. Condoms were easy to find in Europe and were sold unofficially in many government-operated American canteens. When the war ended, venereal disease became a big enough crisis that many Americans began to think of sex as a matter of public health worthy of scientific and social research. The U.S. government began spending millions on public health campaigns to prevent the spread of sexually transmitted diseases. Health officials got help from a judge in the New York Court of Appeals who upheld Sanger's conviction for violating vice laws but established for the first time that doctors could prescribe contraception for women's health.

It wasn't good enough for Sanger. Doctors would only prescribe birth control in the most dire of circumstances, and even then, what form of birth control would they prescribe? There were no reliable options, except perhaps the condom. But condoms depended on the cooperation of men, and Sanger's experience in the tenements of New York City told her that men didn't mind six or seven children so long as they were able to enjoy sex when the mood struck them. Women were the ones dealing most with the consequences of sex, not only because they were the ones getting pregnant but also because they were the ones raising the children. Sanger wrote in 1919:

> *While it is true that he suffers many evils . . . she suffers vastly more. It is she who has the long burden of carrying, bearing and*

rearing the unwanted child. It is she who must watch beside the beds of pain where lie babies who suffer because they have come into overcrowded homes. . . . It is her love life that dies first in the fear of undesired pregnancy. It is her opportunity for self expression that perishes first and most helplessly because of it.

Rubber condoms were sold in tins and had names like Sheik, Ramses, Peacock, and Golden Pheasant. Though they were cheap, effective, and widely available, they were never going to be good enough for Sanger, who cared more about birth control than disease prevention. But the judge's ruling at least gave legitimacy to her cause. She began gaining supporters who once would not have dared to speak out for birth control.

"The Church's attitude on birth control must change," said the Reverend Dr. Karl Reiland, one of the liberal leaders of the Protestant Episcopal Church in New York. "It must support this method of raising the level of existence. Objections on religious grounds are all irrelevant." Rabbi Stephen S. Wise, founder of the American Jewish Congress, argued that "the sacramental attitude toward life does not dictate that there should be illimitable and unchecked generation of life, but that humans shall will to bear children only when they are fitted to give them such a background as will make life worth living." Many doctors agreed as well, and they volunteered their services as Sanger opened more clinics. Even some courts endorsed the legality of contraception. Only Congress, various state legislatures, and the Roman Catholic Church held fast. Importantly, though, the resistance came mostly from entrenched, conservative institutions rather than from the general public. The public was getting Sanger's message, and they were getting it from many different channels.

Sanger grew more sophisticated in her radicalism. Instead of challenging society's conservative views on sex and challenging obscenity laws, she tried to recruit physicians, scientists, and corporate leaders to join her crusade, emphasizing the benefits to public health. She encouraged women to see their doctors to get fitted for diaphragms—not only because diaphragms were the most effective available choice of contraception but also because she knew that doctors would be valuable allies in her fight. A movement centered on sexual pleasure would never get the support it needed, but a movement focused on health might have a chance. It was a strategic accommodation, and a shrewd one.

One of her allies at the time was a wealthy widower named James Noah Henry Slee, who met Sanger and fell quickly in love with her. Slee was twenty years older than Sanger, and a conservative Republican to boot. He was president of the 3-in-One Oil Company, makers of a product that almost every American at the time kept on hand to grease typewriters, bicycle chains, and sewing machines. Sanger had been separated from her husband for seven years when she met Slee. In 1922, she finally divorced William Sanger and married the crusty, aristocratic Slee. There were conditions, though, and Sanger insisted that they be put in writing. They would keep separate apartments, and there would be no dropping by each other's places unannounced. Soon after their wedding, Sanger persuaded Slee to establish the Holland-Rantos Company, the first firm in the nation to sell contraceptives exclusively to the medical community. It was part of her plan to make contraceptives more legitimate. At one point, Sanger asked Slee to spend ten thousand dollars in salary and expenses for a gynecologist to travel the country distributing information about contraceptives. "If I am able to accomplish this," she pleaded with him, "I shall bless my adorable husband, JNH Slee, and retire with him to the garden of love."

Slee went along with Sanger's terms for the marriage. When a drunken friend of Sanger's asked Slee how he could abide such an unusual relationship, Slee replied, "It's obviously an impertinent question, young man, but I'll tell you. She was, and always will be, the greatest adventure in my life."

If it was an adventure he wanted, Sanger gave it to him. And Slee in turn gave Sanger all the money she wanted for her cause, including the money she requested to hire the traveling gynecologist, who went on to present more than seven hundred lectures across the country.

Sanger was making progress. She not only led an advocacy group that operated clinics, she and her husband now owned their own company to design and market birth-control devices. She was approaching the contraceptive market as an industrial magnate might, seizing control of her product from every possible angle. Holland-Rantos gained the loyalty of doctors and became the nation's leading supplier of diaphragms. Doctors liked the arrangement because they were put in the position of supervising women's reproductive health, and they were provided with a new and steady stream of patients. Scientific studies, many of them supported by funding from Sanger, endorsed the Holland-Rantos method of diaphragm and jelly as the safest and most reliable option for women. When the American Medical Association finally backed contraceptives, Holland-Ramos reminded doctors that their contraceptive—referred to as the Koromex method—was the industry standard already used by "over 50,000 physicians, 234 clinics and 140 hospitals."

By now Sanger had broadened her network beyond the original core of radicals, building alliances with powerful people in the mainstream including doctors, wealthy businessmen, and socially active society women. These alliances helped her raise money and helped offset the power leveraged against her by the Catholic Church.

But her most powerful new ally was also her most controversial. Leaders of the eugenics movement were not especially concerned with sexual freedom or women's rights, but they were eager to see certain groups of women produce fewer children, and so they recognized overlap with Sanger's mission. They considered birth control a useful weapon to reduce poverty, crime, and what they referred to as "feeblemindedness." Women most likely to produce poor, feeble-minded, or criminally inclined children should be prescribed birth control—or sterilized—to keep them from reproducing, these eugenicists said. Some of the candidates for sterilization, of course, were the same poverty-stricken women Sanger had set out to rescue when she was a young nurse working on the Lower East Side. Now, however, she found herself acceding to the eugenicists, who enjoyed greater respectability in the United States at the time than birth-control advocates. Eugenics was faddish. In the 1920s, the Harvard psychology professor William McDougall announced that illiteracy could be ended by forbidding people who could read from marrying people who couldn't. Sanger said she saw nothing wrong with using birth control for the weeding out of the "unfit." She agreed that certain people with untreatable, hereditary conditions should be encouraged to undergo sterilization. She also joined calls for criminals, illiterates, prostitutes, and drug addicts to be separated from the rest of society. That these views were widely embraced in the 1920s and 1930s doesn't make them easier to fathom.

Sanger's views on race were complicated. She was a socialist, which irked even some of her strongest supporters, and she was often reckless in pursuit of her goals. But she was not necessarily a racist. In 1930, she opened a family planning clinic in Harlem. The clinic was staffed by a black doctor and had the support of community leaders including W. E .B. Du Bois. Du Bois would later serve on the advisory board for Sanger's so-called Negro Project, which was designed to bring birth control and social services to African Americans in the rural South.

In many ways, Sanger was consistent in her core beliefs. She held that women had the right to self-determination, that every child should be loved and cared for, and that women were entitled to enjoy sex as much as men. In fighting for those principles, she knew and did not mind that things might get messy, as they did in her personal life. Years later, the philosopher Michel Foucault would observe that sex by its very nature was both a private and a social act, something Sanger discovered early on.

※

Sanger's pragmatism and elitism may have damaged her reputation as a crusader for the poor, but, as she expected, it broadened her base of support. By 1925, more than one thousand doctors from around the world sought admission to Sanger's annual birth-control conference, this one held at the Hotel McAlpin in New York City. British economist John Maynard Keynes attended, as did writer Lytton Strachey and Socialist Party leader Norman Thomas. Messages of support arrived from W. E .B. Du Bois, Upton Sinclair, and Bertrand Russell. The most influential eugenicists in the country were on hand, too.

The birth-control movement was gaining visibility in the United States and spreading quickly around the world. It helped that sex in America was more widely discussed than ever. The single woman of the Jazz Age smoked, drank, danced, flirted, and (to use the term that was beginning to come into vogue) screwed. In conservative Muncie, Indiana, home of the classic sociological case study that produced the book *Middletown*, only fifteen of seventy-seven women interviewed in 1924 and 1925 said they disapproved of birth control. And the proof was in the declining birth rates: In the Midwest, the average family size shrunk from 5.4 to 3.3 in one generation. Overall, the birth rate in the United States fell 30 percent between 1895 and 1925, even though women had begun to marry at younger ages.

By 1930, the Birth Control League, the organization Sanger founded, was overseeing fifty-five clinics in twenty-three cities. When critics attacked her for promoting promiscuity, Sanger said she was no more to blame than Henry Ford, whose automobiles made it easier for men and women to slip away to towns where they wouldn't be recognized and commit adultery, often in the backseats of their cars.

Officials in the Catholic Church protested Sanger's assemblies and used their political connections to order police raids. New York's archbishop, Patrick Hayes, said in his Christmas pastoral one year that only God had the right to decide whether children should be born, and that birth control was a more serious sin than abortion. "To take life after its inception is a horrible crime," he said, "but to prevent human life that the Creator is about to bring into being is Satanic. . . . in the latter not only a body but an immortal soul is denied existence in time and in eternity." In some ways, the intervention of the Church was a blessing for Sanger because it nudged her away from issues of sexuality and pleasure and kept her focused on the most powerful argument in her arsenal: human rights. "What he believes concerning the soul after life is based upon theory, and he has a perfect right to that belief," Sanger said in response to Hayes. "But we, who are trying to better humanity fundamentally, believe that a healthy happy human race is more in keeping with the laws of God than disease, misery and poverty perpetuating itself generation after generation."

Sanger was claiming the right for women "to experience their sexuality free of consequence just as men have always done," as Chesler wrote. In her day, that was radical. For the women of Sanger's age, reproduction was the sole purpose for living. Motherhood was the only job that counted. Rare was the woman whose identity extended beyond that of her husband. Even a woman as independent and assertive as Eleanor Roosevelt would be introduced and addressed as "Mrs. Franklin Roosevelt." In attempting to give women the power to rule their own bodies, Sanger was in fact launching a human rights campaign that would have world-changing impact, reshaping everything, including family, politics, and the economy. Once they gained control of their reproductive systems, they would go the next step: They would declare their own identities. Womanhood would no longer mean the same thing as motherhood. Women would delay pregnancy to attend college, travel the world, start jobs, launch magazines, write books, record albums, make movies, or anything else they could imagine. Sanger knew what birth control might do—some of it, anyway. Neither she nor anyone else could have imagined how birth control would also contribute to the spread of divorce, infidelity, single parenthood, abortion, and pornography. Like any revolutionary, she was willing to tolerate a certain degree of chaos.

But she was no longer driven entirely by sex. Now it was sex and money and politics—and soon it would be science, too. Diaphragms and jellies would get her only so far. Even some of Sanger's allies complained that the diaphragm was too expensive and too complicated, especially for poor women. "There is no need to summarize how little we have provided for the people that have the greatest need," wrote Robert Dickinson, a retired obstetrician and gynecologist and consultant to the Margaret Sanger Research Bureau, in a 1931 medical book called *Control of Conception*. "Consider the two and one-half million fertile couples that were on relief. How well can clinics, how well can doctors meet that particular need of the backwoods and the bayou, the requirements of the slum dweller or the distant mountaineer?" He went on to say the only hope was a better form of contraception, concluding: "I propose three ways out. The first is research; the second is research; the third is research." The diaphragm-and-jelly approach would never work in the long run. There were neither enough diaphragms nor doctors to go around.

In 1932, U.S. Customs officials, citing obscenity laws, seized a box of experimental diaphragms sent to Sanger by a Japanese doctor and father of twelve, Sakae Koyama, who believed his new design would make the contraceptives more effective. Sanger and her allies challenged the seizure, arguing that the law was blocking scientific progress and hindering the advancement of medicine. In a landmark decision, a New York State Court of Appeals judge agreed. After that, as long as doctors were involved, it would be legal to use the mail to spread information about contraception or to ship contraceptive devices. That decision opened the door for the American Medical Association to recognize contraception as preventive medicine.

To Sanger, this was real headway. Laws and attitudes were changing. Throughout the 1930s and 1940s, Sanger opened more clinics. During World War II, she promoted "child spacing" as a way to make families healthier and wealthier. The next step was clear.

The field of contraception, she said, "is now ready for a fine piece of research."

SIX

Rabbit Tests

IN JANUARY 1952, Pincus filed a report to Planned Parenthood stating that ten-milligram doses of progesterone administered orally had suppressed ovulation in 90 percent of the rabbits tested. The results were good enough, he said, to justify tests on women, and he was ready to begin.

Pincus wanted Planned Parenthood to think of this as their Manhattan Project. Studies at the time showed one in every four American women had experienced at least one unwanted pregnancy. For Planned Parenthood to seize the opportunity, it would require a huge investment, and the organization's leaders balked. William Vogt was Planned Parenthood's new director, and he was cool to Pincus's work. Vogt, an ecologist and ornithologist, was the author of a controversial and best-selling book called *Road to Survival*, in which he argued that population growth, if unchecked, could destroy the planet. He agreed with Sanger on the importance of birth control, but he lacked confidence in Pincus and the potential use of hormones. Eventually, Vogt wrote back offering to fund him for another year at virtually the same rate as the prior year.

Planned Parenthood's response wasn't a sign that Margaret Sanger had lost interest or given up. It did reflect, though, Sanger's diminishing clout within the organization she helped create. Sanger had many talents, but diplomacy had never been one of them, and by the 1950s even many of the leaders of her own movement were growing frustrated with her. In 1928 she had angrily resigned as president of the American Birth Control League, the organization she had founded, complaining that younger women within the movement were pushing the organization too far into the mainstream. Eleven years later, the American Birth Control League merged with another of Sanger's creations, the Birth Control Clinical Research Bureau, to become the Birth Control Federation of America, which would later become the Planned Parenthood Federation of America. By the time that happened, Sanger had moved to the fringes of the movement. She and her husband were living in Tucson, Arizona, where they had built a home with curved fiberglass walls and stainless steel mantles ("modern as tomorrow," the local newspaper gushed). Sanger served as president of the Tucson Watercolor Guild and hobnobbed with the social elite. When her husband died in 1942 at the age of eighty-three, Sanger inherited five million dollars. She told her son she intended to blow it all and set out quickly to prove she meant it, giving some of the money to the birth-control movement, some to friends, and spending much of it on lavish vacations.

Planned Parenthood had grown rapidly in the 1940s, adding branches across the country. Led mostly by businessmen and male doctors with middle-class women working in the lower ranks, the organization became not only a resource for birth-control services but a wide-ranging health provider for women. Prescott S. Bush, a Connecticut businessman whose son and grandson would both become U.S. presidents, served as treasurer for Planned Parenthood's first national fundraising campaign in 1947. To make the organization's message on birth control palatable, Planned Parenthood stressed issues like "child spacing" and the Asian population crisis. Changing sexual mores were not high on the agenda. Neither was contraceptive research. The leaders of the organization were more focused on building a powerful network, reaching a broader audience, and providing women better access to already available forms of contraception. As the group grew and gained a wider mainstream following, Sanger became a liability at times. She was too loud, too shrill. Some within the organization thought she was politically naïve, and they thought she was especially stubborn and self-destructive in her dealings with the Catholic Church. By then it had become clear that not all of the Catholic faithful agreed with the Church's ban on birth control. Planned Parenthood leaders wanted to try to engage those skeptical Catholics by calling attention to the division between the Church and its members. But Sanger had little tolerance for the Church, and the older she got, the angrier Catholicism made her. What right, she asked, did a bunch of chaste priests have to tell women what to do with their bodies?

Sanger also grew frustrated at times with Planned Parenthood, where officials seemed to be "marking time and holding their own," in Sanger's words. "I'm just as discouraged and discussted [*sic*] as you are—not only about the limitations placed on persons like yourself but the limitations of the caliber of those in charge—to whom you must bow the head and bend the knee," she wrote to Clarence Gamble, one of the movement's supporters, in a 1941 letter she marked "Confidential Please." The relationship had become so strained that she could no longer be certain the organization would fund Pincus's work or any other work she recommended. Proposals would be required. Committees would be consulted. Budgets would be considered. Politics would come into play.

Already she had lost one of her most important financial backers. Throughout most of the first half of the twentieth century, John D. Rockefeller, Jr., had donated millions of dollars to deal with issues of sexual morality. He had not only sponsored the Young Men's and Young Women's Christian Associations to provide havens from the amorality of American cities, he had also funded research into human sexual behavior and donated money for contraceptive research. By the 1940s and 1950s, however, some within the Rockefeller Foundation questioned whether the group's money would be better spent saving lives from malaria and cholera in countries suffering high mortality rates. It was not only a matter of global priorities; there were political sensitivities involved, too. As Alfred Kinsey made headlines with research funded in part by the Rockefellers, foundation officials shied from sex-related projects.

The fault was not entirely Sanger's. In some ways she was a victim of her own success. She had helped bring sex into the mainstream. She had helped build Planned Parenthood into an organization that spanned the country and improved the lives of millions. But as her organization grew, its leaders understandably became less tolerant of risk. They worried about the fate of their organization if Pincus gave them a pill that turned out to have harmful side effects or, worse, killed people or caused women to have children born with deformities. They were frightened by the idea of doing something that had never been tried before: giving medicine to healthy women simply to improve their lifestyles. A scandal or lawsuit could sink the entire organization. Planned Parenthood was not ready to go out on a limb, and certainly not for Pincus.

SEVEN

"I'm a Sexologist"

As a young man, Gregory Pincus fancied himself a poet, philosopher, tiller of the soil, and lover of women. His passion for life and ideas was so great that he sometimes felt it difficult to contain his emotions or to express them in the pages of his diaries. He was tall and handsome with broad shoulders and muscled arms. Even as a boy, there was something uncompromising about his appearance, especially in his eyes. If Margaret Sanger had made sex the guiding principle in her life, arguing that all good things in life flowed from the force of physical love, Pincus had a different core philosophy. "Our one duty," he wrote in his diary as a teenager, "is self-development." Man's job, he went on to explain, is to get the most out of his talents and to help others do the same. In that same teenage diary he tucked away a scrap of lined paper containing a quote from Matthew Arnold, the nineteenth-century poet and cultural critic: "Greatness is a spiritual condition, worthy to excite love, interest, and admiration."

It was not sex, money, power, or fame that guided Pincus. It was that quest for greatness, in Arnold's sense, that drove him. That desire never ebbed as long as he lived.

Pincus was born on April 9, 1903, the first of six children. Genius ran in the family, as did instability. The Pincus family arrived in New

York City in 1891 from Odessa, a cosmopolitan Russian city with a booming economy and many ethnic groups. But anti-Jewish pogroms were sweeping Russia, and the Pincuses fled. Between 1891 and 1910, nearly one million Jews left Russia for the United States, and at times it felt like they were all crowded into an eight-block radius on New York's Lower East Side. This is where the Pincuses settled, briefly.

Gregory's grandfather, Alexander Gregorovich Pincus, opened a restaurant in Manhattan but watched it quickly fail. He moved his family to a farm in Colchester, Connecticut. One day, a man named Hirsch Loeb Sabsovich visited the farm and suggested that Alexander Gregorovich transplant his family to a commune in Woodbine, New Jersey, and enroll his eldest son, Joseph, at the Baron de Hirsch Agricultural School. In the space of a few years, the family had moved from a thriving but dangerous Russian city to a teeming New York City slum to a utopian community of farmers in New Jersey.

The Woodbine Jewish farming colony was a *kibbutz* before *kibbutzes* existed. It was the creation of Baron Maurice de Hirsch, the world's richest Jew, whose family had been bankers to the royal courts of Europe and who believed that the flight from Russia offered his people an unprecedented opportunity to improve themselves and build strong communities around the world. The Baron decided to devote a substantial portion of his wealth to Jewish colonization. He would give away more than $100 million over the course of his life. Some of that money went to the creation of the Woodbine Colony, a 5,300-acre tract in the southernmost part of New Jersey, where the Baron's generosity built houses, schools, barns, factories, a power plant, firehouse, synagogue, theater, and bowling alley. All the streets were named for American presidents except for the main thoroughfare, which was called de Hirsch Avenue. By the time the Pincuses arrived, the population had reached more than 1,400, more than three-fourths of it Jewish. The school had ninety-six students, including Joseph Pincus, son of Alexander Gregorovich.

Though his senior class contained only twelve students, Joseph Pincus was the boy who caught every girl's eye. He looked like a prince, not a farmer, with a long, lean body, elegant features, curled brown hair, and dark brown eyes. "He was so handsome that he took one's breath away," said one Woodbine girl, Lizzie Lipman, the grocer's daughter. Though she came from a family of intellectuals and her brother would go on to become a distinguished professor of agricultural science, women at the time were not encouraged to stay in school. Lizzie had to quit at age fourteen to take a job washing light bulbs at a General Electric factory making three dollars a week, while her brother studied for college. Lizzie was as sharp as barbed wire and just as tough. She begged her parents to give her the same education as her brothers but was denied. "How many nights I cried myself to sleep," she wrote years later, "sick with the realization that sacrifice and service were to be my lot always, that all my ambitions and aspirations were to be stifled and buried in my heart."

Lizzie Lipman yearned not only for an education but also for the handsome young Joseph Pincus, who, after graduating from high school, returned to teach agriculture at the Woodbine school. He was a brilliant man and a strong believer in the use of modern science to improve on nature. He encouraged students and colonists to consider how modern technology might make their plants and animals grow more productively.

While teaching, Joseph fell in love with another instructor. When parents and friends discouraged a marriage between the two, Joseph slipped into a state of depression, an early warning of the emotional difficulties that would dog him all of his life. He left the colony to spend time on a farm in Florida. While he was gone, he and Lizzie exchanged letters. Love unfolded from paper and ink. They married in 1902, and with marriage came the end of their thrilling romance. For Lizzie, like so many women of her generation, a life of sacrifice and service began roughly nine months after her wedding day. In

1903, she gave birth to a son, Gregory Goodwin Pincus, the first of her six children.

In 1908, when Goody was five, the family left the Woodbine Colony and moved to an apartment on Simpson Street in the Bronx near the Seventh Avenue subway, then to Newark, and then back to the Bronx, to a five-story, red-brick apartment building at 741 Jennings Street. The Pincus family joined the Bronx branch of Rabbi Stephen Wise's Free Synagogue, a Reform congregation that encouraged congregants to challenge convention and fight social injustices. Rabbi Wise, an outspoken Zionist, also helped found the National Association for the Advancement of Colored People in 1909, worked on committees to expose corruption in New York City government, and championed the rights of labor unions.

It was while living in the Bronx that Gregory Pincus first experienced sex, at about the age of ten. The family's maid, a Polish woman named Mary, "took him into bed . . . and did things." What things, exactly? He never said. One of his brothers asked him if he'd been raped, and Goody said no, he couldn't have been raped, "because she had a rag around her parts." If he elaborated or spoke about it again in more detail, no one in his family made note of it.

As a teenager, Pincus often used his diary to explore his thoughts and feelings on happiness ("I have been so happy all my life that I can but wish for the same joy in the future"), on religion ("I believe that God, or what I choose to call God, is but the embodiment of all our ideals"), on his flaws ("I really should cultivate a thriftiness and neatness and general economy which I have not heretofore strictly practiced"), on friendship ("the holiest passion on this earth"), and on sex ("My affection easily leads me to express it in a kiss," he wrote in his diary. "My relations I kiss quite without restraint. My friends I cannot kiss at all. How can I express my love for them? It is bursting for expression. . . . I am such an affectionate being. I can't help it").

He had grown into a tall, strong, handsome man like his father, but he cultivated the look of an intellectual, wearing round, wire-rimmed glasses and growing his brown hair long. He complained that his passions were often too intense for others to handle. As an undergraduate at Cornell University, he wrote to his mother to say that while he loved her, he did not share her old-fashioned opinions about sex:

> *Values and standards which have been presented to you as scarcely challengeable I've been foolish (or brave) enough to question. And some of them have not stood up very well. My own conclusions may be very wrong indeed, but they are true for me, and if I were to abandon them, I'd be left floundering, uncertain. Furthermore, I do not think that we differ very strongly in idea. Perhaps as far as sex is concerned there is some slight difference. The sexual impulse is to me neither a low, degrading thing nor something extremely sacrosanct and holy. I regard [it] rather as a fundamentally normal, clean, lifeful instinct. The pleasure derived from its satisfaction, the release of tense emotions isn't sinful of course. But I could never gain any real pleasure from promiscuous satisfaction. I feel that I must love the girl with whom I share this impulse. I do not think that it is necessary for me to wait till marriage for its satisfaction. And I would not feel obliged to wait until I am married because I don't see why a legal ceremony makes it any more beautiful. I think you can understand this. For why should such a natural, fine ardor be repressed and held degraded because a justice of peace has not mumbled a few words!*

Pincus went on in the letter to say he had fallen in love with a young woman named Denah who dropped him because of his "polygamous nature." But he was untroubled. "I am not in love with her anymore, nor she with me. We are both glad of it."

Pincus considered sex beautiful, an expression of love, but he was not as obsessed as Sanger. He did not place sex at the center of his universe and had no intent of making it the subject of his career; he was far too ambitious to consider such a thing. At Cornell, he initially majored in agriculture with a specialty in pomology, the cultivation of fruit. He worked his way through school washing dishes and waiting tables. With his father unemployed, Goody scraped by, often hitchhiking instead of taking the train to go home to his family during vacations. His grades were not exceptional—more Bs than As—and he once was accused of cheating on an exam, a charge that was dismissed after a hearing. But grades weren't everything to him. Pincus sat on rocks and wrote love poems and plays and cofounded a literary journal. He fancied himself a great romantic, yet he also remained every bit the mama's boy, desperate to please.

Goody's mother had six children, five boys and a girl, each one of them smart and ambitious. Lizzie was a confident and intelligent woman who expected greatness from her children—and let them know it repeatedly. "My every thought, feeling, and emotion was given unconditionally to my dear ones, and their happiness was the only reward I asked for," she wrote in a memoir. "And so I go on to the end—hoping, praying, serving, loving—realizing that only the strong are free."

One summer, when his father was yet again unemployed, Goody wrote to his mother to say he was thinking about postponing his studies so he could find a job and earn money. In the undated letter, he carefully laid out all the pros and cons, clearly seeking his mother's approval for whatever he decided. "Mama," he wrote, "we have dreams that we long to see fulfilled. If these are too long deferred they lose their vital meaning, as an underground spring vainly seeking an outlet and finally drying up at the source because it cannot gush forth its sweetness." He went on, quoting Longfellow's poem, *The Village Blacksmith*: "My idea of a successful man is not one who makes a lot of money, or even one who makes a name for himself in the world. Fame is not success. A successful man or woman is one who 'toiling, rejoicing, sorrowing, onward through life . . . goes' with a clear conscience and a big heart."

※

One day during his senior year, Pincus returned home—his family having moved again, this time to Vineland, New Jersey—to find a new occupant, a ravishing older woman, petite, dark-haired, and hazel-eyed, with a prominent bump on the bridge of her prominent nose. Her name was Elizabeth Notkin. After breaking off an engagement to a medical student in Montreal, Lizzie (as everyone called her) had moved to the United States to work as a field social worker with the National Council of Jewish Women. During her training, she boarded at the Pincus home.

The Pincus boys had never seen anyone like Lizzie, who cursed, drank, and chain-smoked Philip Morris cigarettes. She carried herself with the snooty confidence of a New York City intellectual, not the daughter of a mattress factory owner from Montreal.

A cousin recalled meeting Lizzie once and greeting her casually with, "So, what have you been up to, Lizzie?"

"Growing a penis," she deadpanned in her low, gravelly voice, and took a drag of a cigarette.

All the Pincus boys fell in love with her.

Goody was only twenty—about to enter Harvard to pursue a graduate degree in biology—when he encountered this bold and beautiful woman. Lizzie, who was about four years older, treated him at first as if he were a child, asking the young man what he intended to do with his life when he finished school.

Goody, who always loved a challenge, was not to be intimidated.

"I'm a sexologist," he said. Present tense, not future.

This was 1923. No one had ever heard of such a thing. It didn't matter. He had gotten her attention. The next year, when Lizzie was visiting Goody at Harvard, they went before a judge and married without telling their families.

Gregory Pincus was not a sexologist, of course, although he was beginning to read and explore the works of Havelock Ellis and Richard von Krafft-Ebing, the German neuropsychiatrist who published a groundbreaking examination of sexual aberrations in 1886. Harvard, as historian Richard Norton Smith noted, considered itself the "epicenter of American education," a place where a disproportionate number of the world's most intelligent people gathered not only to advance their own intelligence but to launch schemes and dreams that would change the world. Pincus did his graduate work at Harvard's Bussey Institution of Applied Biology, which was founded in 1871 as the university's "agricultural college" but was reformed and expanded to become the nation's leading center of agricultural science. Though Goody Pincus hoped to be among the world changers, he did not get off to a promising start, earning nothing but Bs and Cs in his first year in Cambridge. In the lab, he came under the influence of William Castle, a leading mammalian geneticist, and wrote his dissertation on the inheritance of coat colors in rats.

After his courthouse wedding, Pincus was too busy and too poor to think about finding a home and settling down. So Lizzie moved into the crowded Cambridge apartment Goody shared with his brother Bernard, known to the family as Bun, and one of Goody's friends from high school, Leon Lifschitz, who was enrolled in law school. Goody's mother was horrified by the living arrangements, but, then again, she seemed horrified by almost everything connected with her new daughter-in-law. Goody, Bun, and Lifschitz went in together as partners in a Harvard Square bookshop called The Alcove. Lizzie worked there, too, part-time, although not for long. Soon, like so many newlyweds of her generation, she was pregnant with her first child.

They named the boy Alex John, and they would call him John.

Lizzie suffered through a long and terrible labor giving birth. As a result, she was disinclined to have another child in the years immediately after. What form of birth control she used after the arrival of her son no one knows.

※

When he was twenty-seven and still idealistic enough to believe he could change the world, Goody Pincus was appointed an instructor in the department of general physiology at Harvard. A year later he was promoted to assistant professor of biology, where he worked under a brilliant young physiologist named William J. Crozier, who would both inspire and jeopardize Pincus's career.

Crozier was a hard-driving, ambitious scholar. At thirty-two, he was the youngest associate professor at Harvard. Eager students flocked to him, and he encouraged his protégés to challenge the status quo, to be aggressive, and to use scientific theory to solve practical problems. Crozier was a former student of Jacques Loeb, the German-born American biologist who conducted research in tropisms—or reflexes in which a stimulus causes an organism to orient in a certain way (a plant moving toward light, for example, or roots growing in the direction of gravity). Loeb believed that living things were much like machines, and machines could be made to work the way humans wanted. Taking this theory a step further, he concluded that eggs were the factories that produced these living machines, and if the eggs could be manipulated, life might be artificially produced. Loeb worked with sea urchin eggs surrounded by seawater. When he altered the salt content in the water, the eggs divided and reproduced automatically. Loeb called it *parthenogenesis*—others called it virgin birth—and he promised that it might soon work for mammals. There was no great mystery to life, Loeb suggested; it was all within the control of science.

"I wanted to take life in my hands and *play* with it," Loeb declared, "to start it, stop it, vary it, study it under every condition, to direct it at my will!" The discovery made him a star. Journalists and novelists predicted his work would lead to the factory farming of animals and even children. In 1932, Aldous Huxley published *Brave New World*, a dystopian novel that imagined a future inspired in part by Loeb's work, a future in which people chewed sex-hormone gum to regulate their libidos, women wore "Malthusian belts" to provide universal contraception, and sex was purely recreational. Reproduction by sex was a thing of the past in Huxley's book, replaced by "Hatcheries and Conditioning Centres" that grew babies from test tubes. To some, Loeb was a visionary, a mystic; to others, he was a demon, a scientist playing God, ignoring the unpredictability and beautiful messiness of life. To Pincus, he was a genius whose work hadn't gone far enough. Loeb had never moved on to mammals. Pincus made up his mind to try it himself.

At Harvard, he experimented mostly with rats, studying how they responded to heat and light. When he graduated, he won a fellowship to work on his postdoctorate—studying for two years at Harvard and one year in Europe, dividing time between the University of Cambridge in England and the Kaiser Wilhelm Institute in Berlin. In Europe, for the first time, Pincus began researching the eggs of mammals, the subject that would become his life's work. In Germany, he also observed how scientists were beginning to argue for the use of genetics to create a superior race of humans. As a Jew and a scientist, he was troubled by this misuse of science and criticized it in an unpublished manuscript, arguing that "the race pretension and so called eugenic measure of the Nazi-Fascist variety and genetic nonsense . . . to create a race by breeding and selection would be a gigantic undertaking with a very good probability of failure." When Pincus returned to Harvard as a professor in 1930, he was ready to make his mark and follow in the bold paths of Loeb and Crozier.

While studying and working at Harvard, Pincus met and befriended

Hudson Hoagland, another disciple of Crozier's. Hoagland would later describe Crozier as the leader of "an arrogant bunch of brats." Their research was intended to push boundaries and expand knowledge, regardless of whether it was practical or if it pleased the university hierarchy.

Pincus's interest in how animals passed along genetic traits led him to study the eggs of mammals more deeply—specifically how they were fertilized and how they developed *in vitro* (in test tubes). He tried hormone injections to see how they affected rabbits and at one point noted that estrogen injections prevented pregnancy, but he never pushed on to consider the potential of using such hormones to control human fertility. At the time, Pincus was only trying to better understand how reproduction occurred. He manipulated eggs. He tried fertilizing them outside the rabbits' bodies. He tried transferring eggs from one female to another. He wasn't searching for practical applications so much as playing around and trying to learn what he could.

Within a few years at Harvard, Pincus earned grants from the National Research Council and the Josiah Macy Jr. Foundation. Having chosen an important research subject and made impressive contributions quickly, he seemed poised for a remarkable career. At the time, Jewish researchers at schools such as Harvard had to be better than their peers to earn commensurate respect, and Pincus was doing exactly that. At the same time, though, his benefactors at Harvard were losing power. The university's new president, James B. Conant, disapproved of the way William Crozier had been training biologists like Pincus and B. F. Skinner, who would go on to become one of the most influential psychologists and behaviorists in the world. Soon, the Crozier-led department of physiology was abolished. The timing for Pincus in particular was poor. In 1933, when Conant became president, Pincus's contract was only barely approved.

The following year, Pincus announced to the National Academy of Sciences that he had fertilized rabbit eggs in a test tube, transplanted them into the body of a host mother, and brought the baby rabbits to term. This was typical of the kind of aggressive science that Crozier had encouraged in his students, but it was radical at the time. Crozier wanted his students to recognize the fundamental problems in biology and think about how to solve them. Pincus—the son of a farmer turned teacher who had always pushed his fellow farmers to adopt modern techniques to improve their crops and herds—discovered he had a knack for such work. He wrote in a grant application at the time that his goal was to apply his *in vitro* technique to humans. Almost immediately, his work began to attract an unusual amount of attention beyond the academic community.

It should have been a warning to Pincus when the *New York Times*, citing his work for the first time in a 1934 article, carried the headline, "RABBITS BORN IN GLASS: HALDANE-HUXLEY FANTASY MADE REAL BY HARVARD BIOLOGISTS." The headline was false. Pincus had conceived the rabbits in test tubes and then transferred them to living rabbits, which were sacrificed and studied before the embryos could grow to term. No animals were being born in glass. But it didn't matter. The newspaper went on to portray Pincus as a sinister scientist trying to grow babies in bottles. The article compared Pincus to the fictional biologist Bokanovsky in Huxley's *Brave New World*, who fertilized human eggs in test tubes:

> *At Harvard are two Bokanovskys in the persons of Professors Gregory Pincus and E. V. Enzman [Pincus's collaborator in the study], who have actually taken a step toward realizing the . . . fantasy. Not babies but rabbits have been developed in glass bottles. "We believe that this is the first certain demonstration that mammalian eggs can be fertilized in vitro," Pincus and Enzman remark.*

The *Times* was prescient in at least one respect, explaining that Pincus's technology might eventually emancipate women from the demands of childbirth, separate love from reproduction, and empower the eugenics movement. Pincus claimed in interviews at the time that he was interested only in science, not the implications for humans, but that wasn't true. His work was not theoretical. He was not simply seeking to better understand the workings of sperm and egg. He was engaged, as one writer put it, in "the tinkering of a biological Edison." He was inventing new ways to make babies.

Soon came another story on the same subject, again in the *Times*, "BOTTLES ARE MOTHERS," and another Huxley reference. "*Brave New World*," wrote the *Times*, was mere satire and "may seem amusing to one who knows little of trends in biological research." But when human babies are someday "reared in glass vessels by laboratory experts . . . Gregory Pincus of Harvard is bound to receive his meed of the praise."

After fertilizing rabbit eggs and restoring them to their mothers, Pincus took another step forward: letting the eggs become embryos while remaining in the glass. So far, the newspaper reported, his attempts had failed, but he was getting closer. His new understanding of hormones would be the key. Only recently had scientists discovered that hormones controlled growth, and much of the process remained a mystery. Curious, Pincus planted his fertilized rabbit eggs in a solution that included the hormones estrogen and progesterone. "He expected neither failure nor success," wrote the *Times*. "A good scientist expects nothing; he just watches and draws conclusions." Pincus saw blood vessels forming in the dish after forty-eight hours and a tiny part of the egg that would become a heart. He saw the cells split normally. He counted 128 cells. Instead of a rabbit embryo, though, something irregular and monstrous took shape. It died after fifty-six hours, but such failure was to be expected. Pincus was pleased because the results justified more study. What role, precisely, did the hormones play? Was a mother needed to make them work?

Certain he would soon find answers, he stayed the course. In 1936, Pincus and his colleague Enzman announced that they had achieved parthenogenic development of a rabbit ovum—in other words, they had begun the reproductive process without any fertilization, simply by manipulating the environment surrounding the eggs. Not long after that, Pincus went a step further, saying he had not only achieved parthenogenic development of eggs but had transplanted the eggs successfully into surrogate female rabbits. His "immaculate conceptions," as the press called them, made for more headlines. Newspapers all over the country printed the startling results of Pincus's work, often with headlines like this one: "MANLESS WORLD?" which, not surprisingly, some people found unsettling.

Later the same year, 1936, Harvard marked its three hundredth anniversary with a pamphlet listing the greatest scientific discoveries made by its faculty in three centuries of study. Pincus's work made the list. That same year, he published his groundbreaking book *The Eggs of Mammals*, dedicated to his professors Crozier and Castle, a book that recited the history of man's hunt for eggs:

> *Pfluger, 1863—cat; Schron, 1863—cat and rabbit; Koster, 1868—man; Slawinsky, 1873—man; Wagener, 1879—dog; Van Beneden, 1880—bat; Harz, 1883—mouse, guinea pig, cat; Lange, 1896—mouse; Coert, 1898—rabbit and cat; Amann, 1899—man; Palladino, 1894, 1898—man, bear, dog; Lane-Claypon, 1905–1907—rabbit; Fellner, 1909—man.*

Pincus challenged his fellow biologists to join him in exciting new discoveries that appeared close at hand. "The enormous variety and richness of material that is available and untapped should provide an extraordinary temptation to exploitation now that a beginning has been made in the development of technical facilities for the manipulation of this material," he wrote. "I emphasize that only a beginning has been made." He sounded like a man setting off on a great journey, excited by the possible adventures that lay ahead.

With each new discovery, each audacious claim, and each speech before a scientific body, Pincus attracted more attention in the mainstream press. "The social implications of Dr. Pincus's advance are not easily grasped," said an editorial in the *Times*.

> *Parenthood is still associated much with love. Much of the lyric poetry of the world deals with the wooing of maids, and much of the music and painting that has been given to us by greatest artists are but expressions of the urge that makes grass sprout and lilacs burst forth in the Spring. The more imaginative biologists are not dismayed by looking at a glass vessel and saying: "That's my mother." Serenades will still be strummed on guitars, Romeo and Juliet will still part reluctantly on the balcony. . . . Love will simply be divorced from parenthood if the biologists are right.*

Newspapers all over the country carried reports of Pincus's research. *Time* magazine criticized the scientist for sacrificing so many rabbits merely so their cell divisions could be counted. When reporters asked what his work would mean for humans, Pincus said he was not concerned with such things. He meant that he was interested in pushing science to see where it would take mankind, and that he had no intention of holding back his research because some might be afraid. But that attitude did nothing to calm fears. If anything, it made Pincus sound even more dangerous. And the publicity was about to get worse.

On March 20, 1937, *Collier's* magazine published a feature story on the scientist's work with a strangely lit photo of Pincus, cigarette in his mouth, smoke curling overhead, looking down with dark, hooded eyes at a rabbit held in his arms. If the reader of the magazine got the impression that Pincus's rabbit was not long for the earth, it was understandable, because it was also true. The article began:

> *In the huge Biological Laboratory—a building which represents several of Harvard's fifty-two million dollars' worth of real estate—a 33-year-old scientist leaned over a microscope. His name might have been borrowed from a cop in a detective novel: Gregory Pincus. But what he saw has possibilities more thrilling than anything a detective-story writer ever imagined: a world in which women would be a dominant, self-sufficient entity, able to produce young without the aid of a man.*

The article, perhaps striking anti-Semitic notes, described Pincus as a scientist with "dark penetrating eyes narrowed to slits" and a "heavy mop of black hair." It warned: "The mythical land of the Amazons would then come to life. A world where women could be self-sufficient; a man's value precisely zero." A critic of Pincus's work was quoted in the article saying that if babies were made in test tubes, it "would be the ruin of women." Pregnancy not only improved a woman's looks, this critic noted, but also improved her nervous system.

Suddenly, Pincus was being portrayed as a revolutionary or, worse, a deviant.

Soon after the *Collier's* article appeared, Harvard gave Pincus the news: He would receive a grant to study for one more year at the University of Cambridge in England, and then he would be finished. Harvard was cutting its ties with the young scientist. Pincus believed he was dumped because he had talked too much about his work, particularly to the mainstream press, and because his findings frightened many. No doubt the ouster of Crozier after Conant's arrival hadn't boded well. He wondered if his religion also hurt him. Another factor behind the dismissal may have been Pincus's tendency to publish too soon, as other scientists were unable to replicate the experiments. Then there was the simple fact that he was dealing with sex.

Pincus was thirty-four and had already published a groundbreaking book, as well as a number of attention-grabbing scientific studies. He was on the cusp of what promised to be a brilliant career teaching and conducting research at one of the wealthiest and most prestigious universities in the world. Just like that, it was gone. Pincus may have been the victim of small-mindedness and anti-Semitism, but he was also undone by his own outsized ego.

He scrambled. He applied for jobs but received no offers. He arranged a meeting with Albert Einstein. He asked some of his wealthy and influential cousins for help. But he couldn't find another college willing to hire him.

He appealed to his former classmate, Hudson Hoagland, who had left Harvard and gone to work at Clark University in Worcester, where he had taken over a three-man biology department. Hoagland was a tall, thin man with a bald head, chiseled jaw, and round glasses. Like Pincus, Hoagland saw scientific mysteries everywhere and felt it his calling to solve them. Once, when his wife had a fever, Hoagland drove to the drug store to get her aspirin. He was quick about it, but when he returned, his normally reasonable wife complained angrily that he had been slow as molasses. Hoagland wondered if her fever had distorted her internal clock, so he took her temperature, had her estimate the length of a minute, gave her the aspirin, and continued to have her estimate the minutes as her temperature dropped. When her temperature was back to normal he plotted the logarithm and found it was linear. Later, he continued the study in his laboratory, artificially raising and lowering the temperatures of test subjects until he was certain he was right: higher body temperatures make the body clock go faster, and his wife had not been unjustifiably cranky.

Clark, while not as prestigious as Harvard, was renowned as one of the nation's finest schools for graduate education, which is why in

1909 it had invited Freud and Jung to lecture in observance of the school's twentieth anniversary. It is also why Hoagland thought Pincus might thrive in Worcester. "Knowing his brilliance," Hoagland wrote in an unpublished memoir, "I was incensed that he had not been reappointed and promoted at Harvard and was convinced, rightly or wrongly, that academic politics, including some anti-Semitism, jealousy toward Pincus on the part of some, and antipathy of various colleagues toward Crozier and his group were the reasons for this discontinuance." When Clark said it had no money to hire Pincus, in large part because the Great Depression was straining its finances, Hoagland took matters into his own hands. A rabbi in New York who knew the Pincus family introduced Hoagland to Henry Ittleson, founder of CIT, a holding company that had grown rapidly in the 1920s by financing wholesale suppliers of consumer goods such as automobiles and radios. Ittleson, along with two other donors, agreed to sponsor Pincus's work at Clark for two years. At the time Hoagland had a small grant from G. D. Searle & Co. to study the effects of anticonvulsant drugs in animals, and he asked the pharmaceutical company to transfer it to Pincus, promising that their money would be well spent on this genius's work. Hoagland scraped together enough money for the hire, although Pincus would still be paid far less than other professors at the school and far less than he would have earned at Harvard. He would also be forced to work with a much smaller research budget than he was accustomed to.

But Pincus had no choice. In addition to their son, John, who was twelve, he and Lizzie now had a two-year-old daughter, Laura, as well. Goody's career always came first, and his family would follow wherever he went. The Pincuses arrived in Worcester in the fall of 1938, just as a massive hurricane struck New England, killing almost eight hundred people and destroying tens of thousands of homes. They moved into Hoagland's big house on Downing Street, across the street from Clark's campus, and soon after that into a small apartment two blocks away, where giant tree trunks toppled by the storm remained sprawled across the surrounding yard. Pincus was assigned to work in the basement of the building that housed Hoagland's laboratory, near the bunker where the building's heating coal was stored. But so much coal dust accumulated on his lab equipment that Pincus found it impossible to conduct experiments. He and the other scientists working in the basement were so short of funds they couldn't afford labels for their chemicals. "We had to take the caps off and sniff them to tell which was which," Hoagland recalled. Pincus went back to work on hormones, looking at how they affected the development of eggs.

Still, trouble found him. In April 1939, the Associated Press reported that Pincus had produced two litters of test-tube rabbits from his laboratory in Worcester. "The Clark work was reported by Dr. Gregory Pincus," the AP reporter wrote, "who said emphatically that he is not planning to carry it on to find out whether human babies can be made by test tubes." But in transmitting his story from a scientific conference in Toronto to his office in New York, the AP reporter dropped an important word: "not." As a result, millions of Americans read in their newspapers and heard over their radios that Pincus had stated emphatically that he *did* intend to try to make human babies from test tubes. The AP corrected its error two weeks later, but the damage had been done. Pincus was once again seen as a man with dangerous ideas.

He was operating with no safety net. He still held hope of finding a university job, and he still counted on grant money to do his work. But scandals such as this one didn't help. Much of the government's funding for research in recent years had shifted away from the theoretical to the practical, and so Pincus shifted, too, spending less time on reproductive issues and turning to hormones and how they might be used to help reduce stress in soldiers and make factory workers more productive. In one study, Pincus and Hoagland gave the steroid pregnenolone to laborers in a leather factory to see if it made them work harder and more efficiently, with encouraging preliminary results. Slowly, Pincus rebuilt his reputation. In 1944, a seminal year for his career, he and several other scientists organized a major conference on hormones. They called it the Laurentian Hormone Conference (the meetings were held annually at the Mont Tremblant Lodge in Canada's Laurentian Mountains), and Pincus became the chairman of the conference, a position he would maintain for the rest of his life.

"Everybody is hereby introduced to everybody else," he would say at the opening session of the conference. It was his way of establishing his primacy but at the same time suggesting he expected intimacy to flower among his colleagues. It was classic Pincus.

※

With no teaching responsibilities at Clark, Pincus was free to spend virtually all of his time on research. But he was still treated like a stepchild at the college, unchallenged and underpaid.

In 1940, Goody and Lizzie were raising their two children in a small apartment near the Clark campus. That spring, the principal of John's high school called and said she could no longer keep him in school. He had fulfilled all of his graduation requirements three years early. He had to go. John told his parents he was ready for college and wished to attend Yale. With his parents' blessings he applied—to Yale and nowhere else—and gained admission. He started classes in the fall. The $450 annual tuition (not to mention room, board, and clothing, because John was still growing) put a great financial burden on the Pincuses.

Lizzie grew frustrated at times—not only with the family economics but also with being relegated, as she put it, to the role of "chief cook and bottle washer." At the time, child rearing was depicted in the press and in popular culture as an exciting challenge rather than the endless slate of chores that many women experienced. The so-called "complete woman" was a chef, hostess, storyteller, shopper, gardener, decorator, chauffeur, maid, laundress, and lover. Outside the home, a woman was not expected to have a life. Sometimes when Goody returned home from work, Lizzie would tell him in detail which rooms she had vacuumed and cleaned. In later years, her daughter Laura would wonder if Lizzie honestly expected him to care or if she was only saying it so he would appreciate what an ordeal her home life had become. During World War II, when thousands of women were entering the workforce, Lizzie told her husband she was thinking of looking for work as a radio announcer. She never followed up on it.

Once again, Goody showed no inclination to pursue a steady job. In fact, at the time when his family seemed to be most pressed for money, he was preparing to take his biggest risk. In 1944, he and Hoagland made a move almost entirely unheard of in the American scientific community: They founded their own laboratory, calling it the Worcester Foundation for Experimental Biology. Although it was a challenge to explain to dentists and bowling alley operators that this new foundation would be conducting research on hormones, not curing a dreaded disease, Hoagland and Pincus proved to be excellent salesmen, and the people of Worcester responded generously. Hoagland in particular had a gift for raising money. He came from an affluent family and projected an image of sophistication. Hoagland was also no slouch as a scientist, but it was clear from the start that Pincus was the real genius and Hoagland the organization man.

At first, the scientists operated out of a room at the Worcester State Hospital, but they soon had enough money to hire a dozen workers and purchase a twelve-acre estate in nearby Shrewsbury. Even then, however, money was so tight that Pincus cleaned his own animal cages and Hoagland, shirtless in the warm summer weather, mowed the grass.

The men assembled a guerrilla army of scientists, some of them outcasts, many of them brilliant, all of them lured by the opportunity to work of their own accord, without the pressures of university committees and in the presence of Pincus, whose reputation as a renegade and creative thinker was growing among those smart enough to look past the paranoid associations made by the press. By 1951 the Foundation employed fifty-seven men and women, making it by some accounts the largest privately owned independent scientific research institution in the country. There was no precedent for such an endeavor—one that had no specific goal other than to afford scientists the freedom to explore and invent as they pleased.

Year after year, Pincus and Hoagland made it their practice to spend nearly every dollar raised, putting away almost nothing for emergencies. Ambition fueled them. They wanted every chance to do great research, even as the Foundation building's cesspool—built to accommodate the waste produced by a family, not a scientific laboratory—went long past full and began emitting its contents into the groundwater. Pincus couldn't help himself. When an experiment showed promise, he would hire more scientists and build more lab space to pursue the latest lead, worrying about how to pay later, if at all. A secretary hired in the early 1950s recalled showing up for work and learning that she would have no desk and no office. She got a portable typewriter, a portable table, and a keg of nails for a chair. "Since our general fund has remained low," the Foundation's business manager wrote in a report to the board of trustees in 1950, "it is reasonable to ask whether the Foundation's remarkable growth was either wise or necessary." But Pincus, who made his own office in a converted garage, never paused. He was interested in science and in action, not long-term budgets or endowments.

In those early years, Pincus shuttled his family from one low-rent apartment to another. The eleven-year-old boy who delivered the afternoon newspaper to the scientist's apartment remembered that on weekends he would find Pincus seated in a comfortable chair in the cluttered living room or prone on the sofa napping, always surrounded by towers of books. For one stretch of about six months, while examining mentally ill patients at the Worcester State Hospital, Pincus and his family occupied an apartment in the asylum's main building—the honeymoon suite, as some of the scientists called it. Pincus's daughter Laura would wake up, get dressed for school, say goodbye to her parents, and leave for school, walking past one woman dressed in a flour sack and others obsessively shredding paper and tossing it from windows to make their own snow showers. When asked how she liked living there, Lizzie had responded wryly, "It's like living in a madhouse."

The place had once been known as the Worcester Lunatic Asylum. From the outside, it looked more like a prison than a hospital, a dark and foreboding fortress set high on a hill. Inside, it was worse. Patients were bound by straitjackets and tethered to beds. They underwent electroshock therapy, insulin shock therapy, spinning treatments (in which patients were blindfolded and spun rapidly in suspended chairs), and, under the care of Pincus, hormone therapy. Howls of pain and deranged laughter rang out at all hours, echoing down the tiled halls.

But there was nothing amusing about it. It was a frightening and dangerous place. Why would Pincus expose his family to such conditions? Because it offered a practical solution. It saved money on rent. And it enabled him to concentrate more completely on his work.

※

Pincus did not own a car and did not know how to drive until he was in his forties. He would take a bus or hitch a ride with another scientist to get to work each day. When he did begin driving in the late 1940s, he treated it as a competitive sport, never content to remain behind another vehicle. He was hypercompetitive in other ways, too.

An accomplished Scrabble player, he refused to let his children win and played with a giant dictionary on the table to challenge combinations of letters that might not be words. At the beach, he would swim more than a mile out to sea, calm as could be as his anxious wife and children faded from sight. Even when he read for pleasure—and he devoured mystery novels, especially those written by Agatha Christie and Ngaio Marsh, reading at least a hundred titles a year—he did so as a sort of test, pushing himself to see how quickly he could guess the endings.

He played chess with his insurance agent and met informally to discuss philosophy with a few men who called themselves "The Serious Stinkers." But beyond the world of science and these few recreations, Lizzie Pincus was by far the biggest influence in her husband's life, always pushing him, always making him think, often trying his great reserves of patience. Though she did enjoy cooking and gardening, Lizzie was a far cry from the typical American housewife who stayed home, baked cookies, and greeted her husband each evening with a strand of pearls around her neck, a cocktail in her hand, and a roast in the oven. She usually slept all morning, rising around noon, leaving it to Goody to get the kids washed, dressed, fed, and off to school. Some of her friends and family believed it was a mood disorder that kept her in bed so long and made her so emotionally volatile. Although she could be energetic, witty, and charming, she was also known among friends to undergo rapid and dramatic mood swings.

By the early 1950s, the Worcester Foundation was stable enough that Pincus, after nearly a decade of living in small apartments (not to mention his family's six-month stay at the insane asylum), felt confident enough to buy a house. Given his nature, it was no ordinary home. Located just off the town center in the village of Northborough and wedged between a bank and the town's library, the red brick structure had a dozen bedrooms, ten fireplaces, and a furnished basement where Pincus sometimes offered free lodging to visiting scholars or students. It looked more like an old hotel than a house. Pincus paid $30,000 for it, or about $260,000 in today's dollars. The place was so big that the Pincuses never entered some rooms, as Laura recalled. But Lizzie loved the enormous first-floor living areas, which she decorated in a vaguely Asian theme. The large space, which included a grand piano, was ideal for dinner parties. Lizzie also enjoyed the surrounding grounds, where she planted flowers and vegetables. She was a prolific and talented cook who liked to make big batches of tomatoes Provençal, which she bottled, froze, and handed out among friends. For parties, paid for with Worcester Foundation money, she would cook a smorgasbord of dishes big and small, all done well in advance so that she could enjoy her company. She would circulate, a tumbler of J&B or Johnny Walker scotch in hand, Philip Morris cigarette dangling, working the room until even the driest scientists loosened up and started laughing. Sometimes she plied them with drinks, other times with salty jokes.

"How do you tell if a girl's ticklish?" she would ask, and then pause for effect. "You give her a test-tickle."

The parties lasted late into the night and left nearly everyone drunk. Inebriated or sober, Lizzie was razor sharp, every bit her husband's intellectual equal when conversations were not so scientific. She spoke French and Russian fluently—Goody affectionately called her Lizuska. But not all guests found Lizzie charming. "She reminded me of something of a witch really," one scientist said. "I don't mean she was unpleasant but she had this sort of demeanor." She sometimes lashed out wildly at unsuspecting visitors. Women who paid too much attention to her husband were most likely to come under attack. At other times she would blow up for no apparent reason at all. Laura remembered one explosion that ended with Lizzie storming out of the house and saying she would never return. Calmly, Goody got in his car and drove behind her, creeping along at two miles per hour, until Lizzie relented, climbed into the passenger seat, and came home.

To Goody, the mood swings were familiar. If Lizzie's behavior became too erratic, he would ask if she'd remembered to take her thyroid medicine. But it's unlikely that her abrupt mood swings were the result of a thyroid condition, as Dr. Leon Speroff, Pincus's biographer, has pointed out. By asking about her medicine, Speroff wrote, Goody may have been signaling to his wife that she had crossed a line of acceptable behavior and needed to dial it down. "When she was good she was very, very good," recalled Pincus's brother Alex, "but when she forgot to take her thyroid [medicine] there would be demonstrations of jealousy and temper that through the years alienated a great many of Goody's colleagues, associates, and supporters. . . . [I]t was an important factor in Goody's life history."

Having crashed a car in one of her early attempts at driving, Lizzie refused to get behind the wheel. Goody sometimes asked his secretaries to serve as Lizzie's chauffeurs. Some complied but others refused, in part because it wasn't part of their job descriptions and in part because Mrs. Pincus could be so disagreeable. There was one more consequence of her mood swings and long mornings in bed: Goody, who was colorblind, became known as a poor dresser because he was unable to match his shirts and ties without Lizzie's help.

Through years of marriage to this challenging and thrilling woman, Goody still felt the urge to scribble sappy poems just as he did as a teenager, and on those mornings when the muse struck while Lizzie slept he would leave a few lines of poetry beside her on the pillow so she would read it as soon as she woke.

Even after decades of marriage, Goody and Lizzie remained deeply in love—so much so that some members of the family felt their children did not always get the kind of attention they deserved. Between his long hours at the lab and constant doting on his wife, Goody did not have a lot of time for John and Laura. At one point when Laura was still in grammar school, Lizzie insisted that her husband schedule more time with their daughter. Goody and Laura began making regular trips to Boston to attend plays and concerts. Laura enjoyed the outings, although she remembered her father sleeping through most of the performances.

※

Pincus had no formal office. He had no graduate students to support his lab work. He was both a scientist trying to push the limits of what other scientists had deemed possible *and* the cofounder of a business operation responsible for paying bills and meeting payroll each month. His goal remained the same: to do great work. But it was complicated. And so he zigged and zagged, chased big grant money as well as big ideas, and ordered his staff to continue experiments even when they appeared to be going nowhere. He inspired his subordinates with his confidence and ingenuity.

To his laboratory workers, he was a fatherly figure in an age when fatherly figures were kind and stern but not overly warm. He greeted them with smiles, not hugs or slaps on the back, and he was never seen in the office in anything other than a coat and tie. To new acquaintances and young scientists, he was frightening at times. "He projected the image," said Oscar Hechter, who worked with Pincus at the Foundation, "of a man who had freed himself from trivial matters and was indestructible." M. C. Chang's wife, Isabelle, described Pincus as both charming and terrifying. "I used to be very afraid of him," she said. "When he looked at you, you had the feeling he stared right through you. Nobody dared tell him a lie." At scientific conferences, Pincus made it his habit to sit in the front row, where he would often appear to be sleeping on and off during lectures. But after almost every presentation, he would sit up tall and raise his hand to ask the question that no one else had asked. He wasn't trying to show off or show up the presenter. He asked because he wanted answers. So famous was Pincus for his postlecture questions that at least one scientist said he suffered nightmares preparing for a presentation Pincus planned to attend: "I dreamed that Pincus . . . was sitting in the front row, fingering his mustache and listening to every word, expecting excellence," Sheldon Segal, a colleague, recalled.

Despite his exile from Harvard, Pincus was beginning to establish a reputation for leadership among his peers. He was not only a brilliant scientist, but he also had a gift for organizational work. The Laurentian Hormone Conference became the biggest and most important hormone conference in the world, and as a result, Pincus, with no university or corporate affiliation and no landmark discovery to call his own, became an influential player in the scientific community. He helped decide which scientists would be invited to the conference each year, who would be permitted to present papers, and whose papers would be cited in the yearly conference reports. At the start of each conference, he and Lizzie would host a cocktail party, inviting only about fifty of the conference's hundreds of scientists. "When you went to the Laurentian Hormone Conference," recalled the biochemist Seymour Lieberman, "you genuflected before two people—one was Goody and the other was his wife. . . . She was cocky as hell and she used to keep Pincus in line." To be worthy of an invitation was to know one had truly arrived.

For Pincus, there was yet another reason to volunteer as the conference's organizer: Each year, in reviewing applications from scientists seeking to present papers, he would be among the first to learn of important new advances in the field.

Lieberman was struck by Pincus's Machiavellian tactics and wondered if his aggressive style was a response to the poor treatment he'd received at Harvard. "There were two kinds of people when it came to Pincus," he said, "those who didn't like him and were frightened by him, and those who were just frightened by him. The second group, which was the larger, treated him like an emperor."

Yet in reality, Pincus was an emperor without a kingdom.

EIGHT

The Socialite and the Sex Maniac

IN THE FALL of 1950, shortly before Gregory Pincus first met Margaret Sanger, Sanger received a letter from a seventy-five-year-old woman named Katharine Dexter McCormick. It read:

> *I want to know a) where you think the greatest need of financial support is today for the National Birth Control Movement; and b) what the present prospects are for further Birth Control research, & by research I mean contraceptive research.*
>
> *Sincerely yours,*
> *Katharine Dexter McCormick*
> *(Mrs. Stanley ")*

For Pincus and Sanger, the timing of the letter could not have been more fortuitous. Katharine Dexter McCormick was one of the world's wealthiest women, and after years of personal struggle and tragedy, she was at last free to spend that wealth.

McCormick was the recently widowed wife of Stanley McCormick. In their wedding photo, taken in 1904, Katharine and Stanley posed arm in arm on the grounds of Prangins, Katharine's chateau outside Geneva, Switzerland. The chateau was made of turreted stone and boasted twenty rooms, formal gardens, and a majestic lawn that rolled to the shore of the lake. No one knew exactly how old the place was, but some portions dated to the Crusades. Voltaire had lived there once, as had Napoleon's brother, Joseph Bonaparte.

At the time of her wedding, Katharine was twenty-nine years old, fierce and lovely, a leader in the women's movement and one of the first women to graduate with a degree in science from MIT. She had haunting eyes and a voice so soft and sweet it made men overlook, or at least tolerate, the incendiary things she often said. Deferring her plans to attend medical school, she married Stanley—the youngest son of Cyrus McCormick, inventor and manufacturer of the mechanized reaper and one of the wealthiest men in the world. They were the match of the year, the socialite and the millionaire. Stanley struck an ideal figure: tall, broad-shouldered, athletic, and a Princeton graduate. In their wedding photo, he wears a white tie and tails and clutches a top hat in his left hand. His left knee is slightly bent and his left foot is lifted, as if he's off-balance, not sure what he's supposed to do with the beautiful woman on his arm. In fact, at the time of the photo, he was indeed off-balance, not physically but emotionally. He was hearing voices. Seeing things. The urge to harm women was becoming increasingly difficult to resist. He had been smothered all his life by his powerful mother and now he was marrying an equally powerful young woman, and a frightful storm of emotions was building, catastrophically, behind what appeared in the photo to be a perfectly calm and happy face.

Soon after the wedding, which in so many ways resembled a marriage between two royal houses, the fairy tale turned into a horror show. Katharine may have been attracted by Stanley's timidity and his tendency to give in to her demands, but she could not have imagined how deeply disturbed he was nor what life with her new husband would entail. She began to get a glimpse on their honeymoon, however, when he refused to come to bed and instead stayed awake at night scribbling manically in a book and refusing to tell her what he was writing. One page among the scribbles would turn out to be his will, and in it he would leave his vast fortune to his wife, not his mother.

It was an important breakthrough for Stanley to separate himself from the viselike grip of his mother, but unfortunately the psychic struggle was too much for him to bear. His mind was shattering. Ten months after their wedding, the couple still hadn't consummated their vows, and Stanley's behavior grew increasingly bizarre. Katharine thought he might improve as he settled into marriage and distanced himself from his mother, but the more Katharine talked of her desire for sex and her wish to have babies, the more disturbed Stanley became. Eventually, doctors diagnosed him as schizophrenic.

At the time, most severely mentally ill patients were placed in asylums, where they remained until their deaths, but Katharine McCormick had the money and the determination to do differently. She hired the finest doctors in the world and used her background in science to lead a team of researchers to pursue cures. She moved her husband to Riven Rock, the McCormick family's 34-acre mission-style estate in Santa Barbara, California. Riven Rock had stone bridges, a bell tower, and a nine-hole golf course, as well as heavily padlocked doors and gates. From the terrace, Stanley could scan the grounds, where a lazy creek ran through banks planted with rhododendrons and azaleas, and where he could gaze out at the Pacific Ocean and Channel Islands to the west and the Santa Ynez Mountains to the north and west. It may have been the most gorgeous prison ever built. Ironically, it was Stanley who had built it. Riven Rock had been purchased and turned into a private insane asylum for his older sister, Mary Virginia, in 1897. Stanley supervised its design and construction, never suspecting that his sister would be moved elsewhere and he would so soon become its sole occupant.

Katharine would not divorce her husband, though Stanley's relatives urged her to do so. Even as he became more distant and his behavior worsened, she refused to quit or abandon him. Stanley stuck his hands in the toilet, threw food, and masturbated publicly. He became especially violent in the company of women, which explained why Katharine could only watch him from a distance, sometimes through binoculars, hiding among the begonias in the gardens at Riven Rock. She approved funding to build a primate laboratory—the first in the world—because her husband's doctors believed that by studying apes they would find the cure for Stanley's obsessive sexual behavior. She also dedicated herself to finding a medicinal cure for his illness, putting to use her money, connections, and education in science. She funded research into schizophrenia at the Worcester State Hospital, where Hudson Hoagland and Gregory Pincus would later work, and at Harvard University, launching some of the first programs to examine the link between endocrinology and mental illness. She devoured medical journals, looking for answers that others might have missed.

The study of hormones was new. In fact, the word was coined only in 1905 to describe the chemical messengers secreted by glands and carried to target organs to do a specific job. Katharine McCormick read enough to become convinced that hormones were likely to blame for her husband's condition, and she urged Stanley's doctors to consider hormone treatments. But her access to her husband's doctors—and even to her husband—was limited. The doctors McCormick hired to run Riven Rock, reluctant to take orders even from their employer (in part because she was a woman and in part because they were doctors), wouldn't let her come near her husband, saying that the presence of a woman would send Stanley into uncontrollable paroxysms of violence.

While she supervised her husband's medical care as best she could, Katharine McCormick found herself longing for something more. In 1909 she began volunteering with the women's suffrage movement, speaking at rallies, organizing protests, and providing badly needed funding. Often she was among the oldest women in a room full of young firebrands, but she was wealthy and smart and became a powerful figure in the movement. Mistreated by men at MIT, abandoned by a father who died young, and now all but widowed by her mentally ill husband, Katharine grew increasingly determined to fight for women's rights. She became vice president of the National American Woman Suffrage Association and, after the ratification of the Nineteenth Amendment in 1920, served as the first vice president of the League of Women Voters. At about that time, in the summer of 1921, McCormick began to collaborate with Sanger, who was busy planning the first American Birth Control Conference at the Plaza Hotel in New York. These two powerfully independent women sat at a table as though going over battle plans, papers spread before them, plotting their attack. The birth-control movement intrigued McCormick. She shuddered to think what might have happened if she and Stanley had conceived a child and passed on his sickness. But that wasn't all. Without birth control, any woman might become a prisoner to her husband, a mere breeder. What was the point in fighting for women's rights? What was the point in sending women to college? What was the point in asking women to fight for equality when all they could look forward to was getting pregnant?

Sanger had carried the movement far on her own, but by 1923 McCormick had taken a central role in it. She served on boards, donated money to help publish the *Birth Control Review*, and helped Sanger open the nation's first legal birth-control clinic in Brooklyn. Known as the Clinical Research Bureau, it stayed on the right side of the law by positioning itself as a center for the *study*—not the distribution—of contraception. Of course, the clinic did distribute contraception, and it quickly ran out. Demand far exceeded supply. Diaphragms were routinely smuggled into the country from Canada, but still the Bureau couldn't get enough.

Katharine proposed bringing in more from Europe, and developed a plan for doing it. In May 1923, she sailed across the Atlantic with eight pieces of luggage, including three large trunks. While there, she bought more large trunks, explaining that she intended to purchase many of the "latest fashions" during her trip. She met with diaphragm manufacturers, placed her orders, and had the devices shipped to her chateau. Then she hired local seamstresses to sew the diaphragms into newly purchased clothing, put the clothing on hangers, and packed the exquisite dresses and coats in tissue. Eight large trunks were loaded, sent through customs, and carried aboard the ship as Katharine sailed for home, handing out generous gratuities at every station. Katharine McCormick, aristocrat, smuggler, rebel, arrived at the clinic by taxi trailing a truck containing the most exquisitely packaged diaphragms the world had ever seen—more than a thousand in all, enough to last the clinic a year.

※

In 1927, McCormick offered her chateau as the site of the first international summit on birth control, known as the World Population Conference, organized by Margaret Sanger. But McCormick did not attend. By then, she was caught up in a fight with her husband's family. It started with disagreement about how best to care for Stanley and escalated into a court battle over who should have legal custody of the insane millionaire. "Rich Family Badly Split," read one headline, which was putting it mildly.

Year after year, McCormick gave herself to these two unpleasant and ultimately hopeless tasks, tending to her husband and feuding with his family. She donated money to Planned Parenthood and in 1942 worked with Sanger in an unsuccessful attempt to challenge Massachusetts's restrictive birth-control law. She met occasionally with Sanger to strategize and told the crusading feminist that she believed women would never be free of male domination until they gained control of the reproductive process. She felt a duty to fight for that cause, she said, but there was little she could do so long as most of her energies were devoted to Stanley.

※

On January 19, 1947, at four forty-five in the afternoon, Stanley Robert McCormick died of pneumonia. For more than forty years he had lived in isolation, his mind ravaged by disease, his wife held at a distance, his enormous wealth no help to him whatsoever. Katharine wrote her husband's obituary for the *Santa Barbara News-Press*, noting that her husband's money had been "bestowed generously on many charities and worthwhile institutions" and that the development of the Riven Rock estate had played an important part in the community's economic growth. She went on to note in the obituary that it had cost $115,000 ($1.2 million by today's standards) a year to maintain Riven Rock, and that her husband's medical care had cost another $108,000 a year ($1.1 million). Finally, she pointed out that Mr. McCormick's gardeners had contributed greatly to local horticulture and supported the town's annual Flower Show. The obituary read as if Mrs. McCormick, now seventy-two years old, were attempting to justify the enormous time and money she'd dedicated to what had been, essentially, a lost cause.

When a probative clerk dug through Stanley McCormick's safety deposit box, he found a forty-year-old sheet of hotel stationery, crumpled and yellow. It read: "I hereby bequeath my entire estate to my wife, Katharine Dexter McCormick. I also make her the executrix of the estate." The document had been written and signed on the day of their wedding. Katharine would inherit more than thirty-five million dollars, including almost thirty-two thousand shares in the McCormick-owned company, International Harvester.

Sanger, too, was now a widow, her husband having died in 1943. For Sanger, the death of J. Noah Slee had no great impact. She had felt no passion for Slee in the first place, and so she had little trouble getting on with her life and work. Also, most of her husband's stock and real estate had long ago been transferred to her name.

But for Katharine McCormick, everything changed with Stanley's death. It took her almost five years to settle her husband's estate and come to terms with the Bureau of Internal Revenue, but once those onerous jobs were done, she knew exactly what she wanted to do with her time and money.

NINE

A Shotgun Question

IN JANUARY 1952, before departing on a trip to the Far East, Sanger stopped to visit Katharine McCormick at her mansion in Santa Barbara. Strangers seeing them together on the street might have mistaken them for a pair of wealthy patrons of the arts. McCormick was still an imposing figure, tall and lavishly dressed, even if her fashion sense seemed to have frozen somewhere around 1930. Sanger was the shorter, lumpier of the two, a copper-haired, blunt-talking bundle of energy. The women spent much of their time discussing their health and diets. But inevitably the conversation also turned to sex.

In a letter written around the time of the meeting, McCormick said she was "feeling pretty desperate over the research end of our work." In the past, Sanger had tried to persuade her wealthy friend to donate money to Planned Parenthood. But she wasn't sure that was the best approach anymore.

As Sanger was preparing to leave for Asia, Pincus sent a report on his progress, describing in four single-spaced pages the experiments he and Chang had performed on rabbits and rats and explaining the effects of hormone injections versus administration by mouth. He confirmed that the oral doses were only 90 percent effective and said he hoped to experiment with different progesterone compounds that might work better.

"The foregoing experiments," he concluded, "demonstrate unequivocally that it is possible to inhibit ovulation in the rabbit and successful breeding in the rat with progesterone. . . . It has been demonstrated furthermore that following the sterile period, normal reproduction may ensue." The thing to do next, he said, was to try more progesterone compounds.

In response to his report, Pincus was grilled by Planned Parenthood's national director, William Vogt, who wanted to know where Pincus's work was going to lead: "In what specific ways," Vogt wrote, "can you anticipate—if you will attempt such a projection—that the results of your research may be put to work? That is a shotgun question, but I believe valid, because a constant problem in most research is selling the idea of research itself, especially to those who may help support it."

Pincus was getting drilled for answers after only a year of work and an investment, including funds from McCormick and Planned Parenthood, that had amounted to a mere $3,100 ($27,000 today). For 1952, he was expecting to receive $3,400. He wasn't complaining, but he and Sanger both recognized that Planned Parenthood had not yet made a serious commitment. The organization's leaders, Sanger told McCormick, "have evidently not been sold on the Pincus research." She suggested to McCormick that they might have to take matters into their own hands.

Soon after, in June 1952, McCormick made plans to visit the Worcester Foundation to see firsthand what was happening there. She met Hoagland and Chang and learned about the progesterone work underway, but she did not see Pincus, who was out of town on other business.

※

One weekend that fall, Pincus and twenty-nine scientists from eighteen cities and twenty universities met at Arden House, a conference center in Harriman, New York, overlooking the Ramapo Valley. It took Pincus's enormous influence to bring together such a distinguished group of biochemists, gynecologists, endocrinologists, immunologists, and sociologists, and to do it over a weekend, no less. But given that the meeting concerned a sensitive subject, it's possible that the scientists preferred to do such work on weekends and away from their universities. No publicity was sought for the gathering, and the names of those assembled were omitted from the documents produced in the meetings. The objective of the assembly: to discuss the expansion of research and testing of fertility control.

In a report summarizing the discussions held during the meeting, the scientists agreed that most of the progress in birth-control research had been the accidental byproduct of other work. They also agreed that "safe, effective, inexpensive and aesthetically acceptable" oral contraceptives were well within reach—if only researchers would commit to their development. They went so far as to issue a resolution:

> *As scientists and individuals, we wish to be on record as recognizing the importance of expanding research directed toward improvement and more widespread application of acceptable means of human fertility control. Based on our understanding of trends in population growth and resource utilization, and on our belief that research on fertility control will contribute significantly to the alleviation of basic problems underlying world tensions, we endorse this recognition with a strong sense of urgency.*

Powerful words—except that no names were attached. If he didn't know it already, Pincus could clearly see that there would be no groundswell of support from the scientific community.

※

"I am rather surprised that the Pincus plan does not receive more attention," McCormick wrote to Sanger on October 1, 1952. "Perhaps it is considered too long-ranged and complicated—but I wonder if it is any more so than the others." She reminded Sanger, who should have needed no reminding after nearly half a century of work, that a birth-control pill was not likely to be a simple invention, and that investors and organizations backing it would have to be patient. With her degree in science from MIT, and based on her visit to the Worcester Foundation months earlier, McCormick had become Sanger's leading expert on the Pincus plan for contraceptive research. Or perhaps Sanger was merely pretending to be naïve in the hopes that her wealthy friend would step into a position of leadership.

"It is pretty trying not to be able to *push it*!" McCormick wrote to Sanger in another letter.

Sanger sympathized. "You are quite right in feeling as you do about research," she wrote. "In a few months, perhaps we can go see Dr. Pincus."

TEN

Rock's Rebound

FOR YEARS GREGORY Pincus had been searching for a project that might establish his greatness, only to watch ideas come and go like love affairs, beginning with promise and ending in hurt feelings. His whole career thus far had been a recovery process, one attempt after another to start over. By now, he was smart enough to know that the quest for an oral contraceptive carried enormous risk. The pill could fail; it could cause serious side effects; it could stir caustic publicity and once again make him a pariah.

Despite the perils, however, the project was perfect for Pincus. It concerned the area of science he knew best: mammalian reproduction. It called for a scientist trained to think aggressively. And it required not only scientific knowledge but also an entrepreneurial spirit, which Pincus had developed since he'd been forced to leave Harvard. But the best reason the project suited Pincus was this: He had nothing to lose; his participation in a controversial project such as this one was unlikely to harm his reputation or professional standing any more than it had already been harmed. As one of his colleagues put it, "He wasn't afraid to go out on a limb because he didn't have any limb."

Years of disappointment had taught Pincus that it wasn't always the science that determined an experiment's success; it was often the countless forces surrounding the science, some of which were within his control and some of which were not. What the pill project needed most now was not necessarily a biologist but a product champion—someone who could build the team to do the scientific work, forge alliances with manufacturers needed to supply chemicals, and, if all went well, spread the news of the coming invention so that it might have a chance at acceptance. Had Pincus still been on the faculty at Harvard, it's unlikely he would have had the drug company connections that he cultivated on his own. Nor is it likely he would have had the high tolerance for risk that he did. In the years ahead, he would not only have to risk his reputation, he would also have to push the boundaries of law and ethics.

He knew what he had to do next: test progesterone on women. And to do that, he would have to add a player to his team—a doctor, preferably a gynecologist, someone who could reassure the patients involved in the experiments that they were safe and would convey to the drug companies supplying the progesterone that no one would be harmed by the experiments. He considered Dr. Abraham Stone, who was sixty-three years old and one of the nation's leading experts on contraception. It was at Stone's party in New York in the winter of 1950 that Pincus had first met Sanger and begun thinking about how a pill might be formulated. Sanger certainly would have approved of the choice, but Pincus worried that Stone, director of the Margaret Sanger Research Bureau, was too closely aligned with the cause to be objective. He would likely be seen as an advocate for birth control, giving the science a partisan tinge. He also considered Alan Guttmacher, chief of obstetrics at Mount Sinai Hospital in New York, author of a popular marriage manual, and advisor to the Population Council. But Pincus worried that Guttmacher would be too busy to give the progesterone experiments his full attention. There was one more concern: Both Stone and Guttmacher were Jewish. Pincus, still feeling like he'd been stung by anti-Semitism at Harvard, worried that having another Jew on the team might invite criticism.

Pincus's next choice was a physician named John Rock. Like Pincus, Rock was a Harvard man. Rock was respected by his peers and adored by his patients. He looked like a family physician from central casting in Hollywood: tall, slender, and silver-haired, with a gentle smile and a calm, deliberate manner. Even his name connoted strength, solidity, and reliability.

Rock had one more thing going for him: He was Catholic.

※

In 1890, John Rock was christened at the Church of the Immaculate Conception in Marlborough, Massachusetts. As a teenager, he was so devout in his faith that he felt compelled to confess to his priest every time he felt a sexual urge or experienced an erection. He even wrote down the dates and the number of times of these occurrences until, at last, in the confessional booth, his priest told him: "Don't be so scrupulous, John."

He was the son of an Irish saloonkeeper. Though he was big, strong, and athletic, he often preferred playing with his sisters in the house rather than mixing it up on the streets or in the yard with his brothers. For this, his brothers sometimes called him a sissy. In the spring of 1907, while enrolled at the High School of Commerce in Boston and living in a boardinghouse with his fellow students, young John Rock may have fallen in love with a classmate named Ray Williams, who was the captain of the school's basketball team. Rock's diary from 1907 is filled with excited scribbling about his time spent with Ray. In March of that year, however, something happened that prompted Rock to tear out pages from his diary. Sixty-five years later, as an old man reflecting on the dawn of his sexual conscience and his liberal views on sex and contraception, Rock mentioned sleeping in the same bed with a friend he refers to as *Ben* Williams and waking up with an erection and, summarily, an orgasm. In later years, Rock would show remarkable open-mindedness about varieties of sexual behavior. His friendship with Ray might have helped shape his thinking.

After high school, Rock surprised his working-class family by gaining admission to Harvard and studying to become a doctor. In 1926, he became director of the sterility clinic at the Free Hospital for Women in Boston. He loved his work and adored his patients. On busy days, he would hustle between two exam rooms, trying to see as many patients as possible without keeping anyone waiting too long. He would often ask his poorer patients if they needed bus fare home, and if the women were extremely uncomfortable in their pregnancies he would walk them out to a taxi stand and pay the driver in advance. In his carefully tailored suits, crisp shirts, and handsome ties (he seldom left the house without a tie or ascot), he was the picture of formal elegance. His clothes were not expensive. Rock would never betray his working-class roots, nor would he dress so ostentatiously as to intimidate his patients. But he insisted on conservative attire at all times when seen in public, and he was equally prescribed in manner. He worked with one nurse for twenty years without ever learning her first name; she was always "Mrs. Baxter" and he was "Dr. Rock."

Though he hung a crucifix over his desk at work, Rock's religious and professional opinions often conflicted. While the Catholic Church opposed abortion, for example, Rock believed a woman's health was more important than the health of her fetus and that pregnancies should be terminated when they imperiled patients' lives. "Religion," he used to tell his daughter, "is a very poor scientist."

In 1936, while serving on the American Medical Association's Committee on Maternal Health, he told fellow committee members he believed that sex was meant for reproduction, nothing more. "Nature intended motherhood to be woman's career," Rock said. Anything that diverted a woman from starting that career immediately upon marriage, he continued, "is socially wrong." He'd heard of cases in which women wanted to postpone pregnancy so they could earn money to help their husbands finish their college degrees, but he had little patience for such wishes. Let the man postpone his education so the wife can have her baby, he said. As for sex, it could never "be made an end in itself without dire consequences."

Over time, however, Rock underwent a fundamental change as compassion for his patients overwhelmed his compulsion to toe the Church's line. He sympathized with women who came to his practice saying they were afraid of becoming pregnant again, whether it was because their bodies were worn out or because they couldn't imagine caring for more children. Rock also began to see that many couples wanted contraception because they wanted to delay, not avoid, becoming parents. In 1931, he was one of fifteen Boston doctors (and the only Catholic) to sign a petition calling for the repeal of the state's ban on contraception.

In 1925, Rock married Anna Thorndike, a Boston woman who shared his sense of adventure, having served as an ambulance driver in France during World War I. The Rocks married relatively late in life—he was thirty-five and she was twenty-nine. They had their first child eleven months after their wedding. Four more children came in the next six years. And then they stopped coming. Rock never discussed why or how.

He adored his wife and, unlike many men of his era, had no fear of showing his affection publicly with flowers, eloquent speeches, and stolen kisses in the hallway of their home. In his medical practice, Rock counseled pregnant women and delivered babies, but he also worked with women who could not conceive and came to believe that sexual intercourse offered an important bond for husbands and wives as they struggled unsuccessfully to have children. Too many priests, he said, confused the beauty of human coupling with animalistic copulation. Their obstinacy frustrated Rock and caused him to state his views on sex and love more boldly. By the 1950s he began to lecture on the subject, saying sex and love were inseparable. Only with love, he declared, could "orgasm reach its natural fullness of ecstasy."

That was not exactly what he'd been taught by the priests.

※

There is no mention of contraception in the Bible, Old Testament or New, nor did the term enter the vocabulary of Catholic moral theology until the second half of the twentieth century. Before then, the most relevant term used by theologians was *onanisma*, from the biblical story of Onan (Genesis 38:4–10), which was described as masturbation or sexual intercourse performed without the intention of reproduction. Sex was only for procreation, the Christian church declared, which made *onanisma* a sin.

The human reproductive system was poorly understood even in the early years of the twentieth century. Many people thought women were merely the vessels, and that it was the man's seed that sprung on its own into a baby. That's why spilling seed, or losing semen, whether in sex or masturbation, was labeled a sin. The philosopher and theologian Thomas Aquinas wrote extensively and influentially on the subject in the thirteenth century, arguing that all sex without procreation, even within a marriage, amounted to lust. Still, the Catholic Church had no official position on birth control until 1930, when Pope Pius XI issued a papal encyclical called "Casti Connubii" (Latin for "Of Chaste Wedlock"). The pope acknowledged that birth control was widely used "even amongst the faithful," although he wasn't happy about it, and called this trend "a new and utterly perverse morality." He added that it amounted to a "shameful and intrinsically vicious" attempt to get around the natural "power and purpose" of the conjugal act. The pope did, however, offer the faithful an important loophole: A married couple would not be sinning, he said, if for "natural reasons either of time or of certain defects, new life cannot be brought forth." In other words, a married couple could have sex for pleasure so long as the husband and wife knew that natural reasons prevented them from having children.

For decades doctors had been instructing women who did not wish to become pregnant to have sex only during their "safe periods." Unfortunately for many women, until the 1930s most doctors believed the safe period came in the middle of the menstrual cycle; in fact, that's the time when women are most likely to conceive. After scientists finally got it right, a Chicago family doctor named Leo J. Latz, a devout Roman Catholic, figured out how this information, combined with the pope's recent declaration, offered men and women a shot at having guilt-free and baby-free sex at certain times of the month.

Latz wrote a dryly worded book called *The Rhythm of Sterility and Fertility in Women*, which sold hundreds of thousands of copies. By avoiding intercourse for eight days each month—five days before ovulation and three days after—women could naturally and ethically control their bodies and regulate their family sizes, Latz instructed. It wasn't foolproof, of course. Timing ovulation is tricky. Every woman's body is different, and a woman can ovulate at a different time each month, depending on factors such as stress and illness. But Latz did his best to help. He told women to keep a detailed record for six months of the exact dates when their menstruation began and then count the days in between. Once they discerned a regular cycle, women could determine their infertile days based on research that showed ovulation occurred twelve to sixteen days before menstruation.

Despite the uncertainties, Latz's message found a receptive audience, in part because he wrote with one clear assumption in mind: that married couples had a right to enjoy regular, fear-free sex simply for the pleasure it provided. God planned it that way.

But there was more than pleasure on the line. Women all over the world were desperate to control family size or better time the arrival of children—for the sake of their health and the welfare of their other children. Companies began producing graphs, wheels, calendars, and slide rules to help women accurately calculate their cycles. Despite the great impact of *The Rhythm*, however, it eventually cost Latz, who was fired from the medical faculty of Loyola University in Chicago, almost certainly because of his controversial cause.

In the 1930s, birth rates for all American families fell to a low of 2.1 children per mother, in large part because of the Great Depression. Catholic families remained larger than average, but even Catholic couples began having fewer children as women became increasingly comfortable with the rhythm method and other forms of birth control. "The trend of our Catholic population is toward extinction," said Monsignor John A. Ryan in 1934. "Our people are showing that they have not the capacity, the courage and the endurance necessary to marry and bring into the world sufficiently large families to ensure group survival." Many priests took to their pulpits to attack birth control and abortion, but their sermons did little good. Birth control was out in the open now. For the first time, many Catholics began compartmentalizing their beliefs. Sex was something private and apart from religion. They would obey the pope on *this* matter but not *that* one. It was the rumbling before a seismic shift.

Margaret Sanger should have been satisfied that birth control had driven a wedge between the Catholic Church and many of its followers. She should have been happy that so many Catholic women were coming around to her way of thinking. But the rhythm method wasn't good enough for her. It was unreliable, which is why the joke was born: "What do you call a woman who uses the rhythm method? *Mommy*." It also did little to help a woman explore her sexual desires, since it required women to limit those desires to certain times of the month. Sanger still wanted doctors to dispense a reliable, low-cost form of birth control. And she did not want unmarried women to be excluded. Partial victories were not victories at all in her mind. Her Catholic critics complained that the rhythm method was superior to Sanger's artificial contraceptives because it did not interfere with the natural process of life. But Sanger shot back that all sorts of things interfered with the natural process of life. Resisting sexual temptation interfered with the natural process of life. Every time the pope shaved his whiskers, she said, he interfered with the natural process of life.

Sanger's mistrust of the Church had grown and calcified over time. So great was her distrust that she did not want Pincus to include Rock on the team investigating progesterone, arguing that "he would not dare advance the cause of contraceptive research and remain a Catholic." Pincus defended Rock, saying that the doctor was a "reformed Catholic" whose medical views were distinct from his religious beliefs.

Sanger did not lose a lot of arguments, but she lost this one.

※

Pincus saw in Rock not only a talented scientist but also an important promoter of his new, as yet unrealized birth-control pill. Rock had already gained a small measure of fame as the Catholic doctor who dared defy his own church. In 1944, he made headlines when he and his assistant, a former Pincus lab worker, achieved the first successful *in vitro* fertilization of human ova. Rock neither bragged nor attempted to frighten, telling reporters it would take at least a decade before the technology would evolve enough to produce pregnancies for women. Unlike Pincus, the urbane, pipe-smoking, ascot-wearing Dr. Rock offered a reassuring presence. No one would have dared compare him to Dr. Frankenstein. Quite the contrary, newspaper and magazine readers looking at his picture and taking in his measured words couldn't help but feel the future was in good hands. If John Rock said it would be safe, they concluded, it must be safe.

Even when Rock challenged the Catholic Church, he did so diplomatically, giving the impression that he was on the side of fairness and tolerance. He wasn't trying to blow up the institution of marriage. He wasn't trying to encourage sex for the sake of mere pleasure.

He wasn't trying to hurt the Church. He was encouraging Americans in general and Catholics in particular to be more thoughtful in how they approached marriage and the making of families. "I don't think that Roman Catholicism forces a man to interfere with other people's freedom of conscience and action within their own moral principles," he told *Time* magazine in 1948. Soon after the publication of the *Time* article, Rock published a book called *Voluntary Parenthood*, which attracted more national attention, including an excerpt in *Coronet* magazine, a general-interest magazine that rivaled *Reader's Digest*. "Nothing in the life of a man and woman is going to be as important to themselves or society as their parenthood," Rock's article began. "It seems reasonable, then, that prospective parents should apply at least as much intelligence and foresight to this as to designing a home, buying furniture, or choosing a career. Yet, when a modern couple embark upon the voyage of life together, they are launched on a sea of ignorance. . . . They have to find their own course, for the available charts are mostly a mixture of superstition, science, and symbolism."

Rock, in his own way, made it his mission to improve the available charts. He wanted young couples to talk about sex and babies before they married. He wanted them to understand that sex was neither shameful nor obscene. He wanted society to provide safe and effective means of birth control, and he wanted married couples to have the right to use them. For all of this, Monsignor Francis W. Carney of Cleveland called Rock a "moral rapist," and Frederick Good, the head of obstetrics at Boston City Hospital, asked Boston's Cardinal Richard Cushing to have Rock excommunicated. But Rock would not budge. It was no wonder Pincus was drawn to him.

※

Gregory Pincus and John Rock first met in the 1930s, when Pincus was still at Harvard. In the 1940s, when Rock began experimenting with *in vitro* fertilization of human ova, one of his first steps was to send his research assistant to Pincus for guidance.

When Rock treated women for infertility, he would begin by taking a medical history and providing a complete physical exam. If the woman wasn't menstruating, or if she wasn't menstruating regularly, Rock would suspect an ovulation disorder and order an endometrial biopsy. Rock was unusual among fertility specialists at the time because he also asked husbands to have their semen tested. He suspected (and his suspicions would be confirmed in later years) that men were responsible for a large percentage of infertility problems. He was also unusual—if not unique—in that he operated a rhythm clinic down the hall from his infertility clinic.

Rock's rhythm clinic was the first free clinic in Massachusetts to offer birth-control advice. After receiving physical exams, women visiting the rhythm clinic were asked to chart their menstrual cycles and sex lives for three months. After that, Rock tried to instruct the women who had regular cycles when they could safely have sex with little risk of conception. He knew that many of the women were using diaphragms, douches, and condoms, but the law would not allow him to prescribe or even discuss those items unless the woman's health was in serious jeopardy. Even if Rock had been allowed to distribute birth-control devices, roughly 90 percent of his patients were Catholic and more interested in the rhythm method. Some of his patients had had more than a dozen children and wanted to stop getting pregnant. Others wanted to better space the arrivals of future children. Between the women seeking birth control and those patients who were trying to overcome infertility, Rock came to understand not only human reproduction but also a good deal about human relations. In the same day, he would see some women who were straining to raise more children than they could handle and others deeply wounded by their inability to get pregnant. They were housewives, mostly working-class women, married to bakers, laundry workers, elevator operators, and machinists. Among the women with children, many came asking for the only thing they'd ever heard of that would guarantee an end to their baby-making days: a hysterectomy, or the removal of the uterus.

One such patient, known as Mrs. L. A., was thirty-two years old. She had married when she was eighteen, borne eleven children, and had one miscarriage. Her last five deliveries had been by Cesarean section, and her very last had been twins. She told Rock she and her husband had sex twice a month and never used birth control. The twins were only six months old when Mrs. L. A. visited Rock. She reported that her husband was trying to be "careful," meaning that he was withdrawing before he ejaculated, to avoid getting her pregnant again. When Mrs. L. A. arrived at Rock's office, she was sent to the fertility clinic accidentally. Once that was straightened out, she told the doctor she was exhausted, in pain, and suffering occasional blackouts. Her periods were unusually profuse and painful, she said. Rock, concerned that the woman might have a tumor, suggested an immediate hysterectomy.

Another patient, Mrs. M. B., regularly used contraception. But even so, after eleven years of marriage she had six children (and had suffered one miscarriage) before she turned thirty. She visited Rock after a botched abortion. The first time Rock saw her at the hospital, he refused to approve a hysterectomy, which clearly would have been intended to sterilize her. Instead, he suggested that she get fitted for a diaphragm. When she returned to the hospital yet again after having become pregnant and attempted to perform her own abortion, Rock relented and agreed that a hysterectomy was appropriate.

Meeting these women emboldened Rock. He had long believed that the Church and the state of Massachusetts were wrong for rejecting birth control, but he was gradually becoming radicalized. In 1945, he wrote to one hundred women to ask a series of questions about how their hysterectomies had affected their health and well-being.

He wanted to know how the operation affected their marriages and sex lives. About half of the women told Rock their sex lives were no different from before the operations. Five said their sex lives were worse. Eleven (including one woman who had dumped her husband for a new man) said their sex lives were improved. Three women said their husbands had rejected them since the hysterectomies because, as one of them put it, she had lost her "nature."

Not many doctors had the courage or confidence to ask women about the quality of their sex lives. The more Rock got to know these women, the more interested he became in birth control. In the 1930s and 1940s, most of his work—and the work that interested him most—centered on helping women overcome infertility. At the same time, he was becoming the most famous fertility specialist in the country. The rich, powerful, and famous traveled from across the country to see him, among them Merle Oberon, a native of Bombay who starred in Hollywood features that included *Wuthering Heights* and *The Dark Angel*. At the age of sixteen or seventeen, Merle had been sterilized. Her mother, hoping to protect her beautiful daughter from an early pregnancy, had ordered the procedure without explaining its implications to the young girl. By the time she was in her thirties, Oberon wanted children and asked Rock if it was possible to reverse the procedure her mother had forced on her.

In the 1950s, every young adult woman seemed to be having children, or wanting to. Raising children was an act of patriotism in postwar America. It was the key to happiness, the path to fulfillment. Men and women who couldn't reproduce were pitied. Year after year in the 1950s, the nation became more fertile. In the 1930s, the average American woman had about 2.2 children over the course of her lifetime. By 1957, that number would hit an all-time high of 3.7. Women who weren't getting in on the baby-making action were considered incomplete and undesirable. Men were shamed for "shooting blanks" and had their masculinity questioned. Since doctors knew little about what caused infertility, women took most of the blame. Usually, physicians pointed to psychological causes, saying the women were suffering stress or else they were subconsciously afraid of having children. In 1951, an article published by a sociologist, gynecologist, and psychologist (all of them men) in the *Journal of the American Medical Association* pronounced that women who lacked the desire for children were so rare "they may be considered as deviants."

Even Dr. Abraham Stone of Planned Parenthood argued that infertile women needed to adjust their attitudes if they wanted to get pregnant: "For conception to take place a woman must be a woman. Not only must she have the physical structure and hormones of a woman, but she must feel she is a woman and accept it. . . . Being a woman means acceptance of her primary role, that of conceiving and bearing a child. Every women has a basic urge and need to produce a child."

Demand for fertility treatments exploded in the 1950s, but doctors offered little help. Beginning around 1950, Rock conducted a series of experiments on women struggling with what he called "unexplained infertility." He suspected that some of the women were not conceiving because their reproductive systems were not fully developed. When a woman with such a condition did somehow become pregnant, the ensuing pregnancy helped her reproductive system mature. To test his theory, he recruited eighty "frustrated, but valiantly adventuresome" women for an experiment in which he would use hormones—progesterone and estrogen—to create "pseudo pregnancies." He confessed to the women that he had no idea if it would work, but the women trusted him and went along.

Rock's experiment was based on his understanding of three kinds of hormones:

- Androgens, the masculine hormones, although they are present in both sexes. The most active androgen is testosterone. These hormones control the development and function of the male genitals. They promote a man's muscular strength, the growth of his facial hair, and his deep voice. In both men and women the androgens help account for sex drive.

- Estrogens, the feminine hormones, originating mainly in the ovaries. Estrogens promote the development of breasts, the lining of the vagina, and the lining of the womb in preparation for pregnancy.
- Progestogens, often referred to as the pregnancy hormones—with progesterone being the most influential in the category—because they regulate the condition of the inner lining of the uterus. When an egg is fertilized, progesterone prepares the uterus for implantation and shuts down the ovaries so no more eggs will be ovulated.

In 1952, Pincus and Rock attended the same scientific conference and chatted between sessions about their work. When Rock described his work with the pregnant women, Pincus suggested to the doctor that he try progesterone without estrogen.

Rock experimented with different combinations of hormones and different methods of delivery, trying to find the medicine that worked best and caused the fewest side effects. Sometimes, before giving the hormones to his patients, he injected himself to assess how much pain each dose packed and, presumably, to make sure he wouldn't drop dead. But beyond that, he did little in the way of preparation. He did not test the hormones on rats or rabbits. He did not ask his patients to sign consent forms, but he did explain to the women that the medicine they were about to take would not directly help their infertility. The knowledge he would gain from the experiments might someday help a great many women, he explained, but he was careful not to make promises. This was how experiments were done at the time, and John Rock's word was enough for most of his patients. "Like us," he said, "they wanted to try it."

He started the women on fifty milligrams of progesterone and five milligrams of estrogen and escalated gradually to three hundred milligrams of progesterone and thirty milligrams of estrogen. When the first round of treatments ended, no one was dead and no one had become seriously ill. That was good news. Within months, the news got better. Thirteen of the eighty women in Rock's care became pregnant. Rock told his colleagues about this promising result. "The rebound reaction," he called it, because the hormone-induced shut-down of the women's reproductive systems seemed to have given their bodies a lift and helped them become fertile. Soon, his fellow gyne-cologists were calling it "The Rock Rebound."

Rock was encouraged, but he wasn't entirely convinced the rebound was real. The sample size was too small, and the unexplained infer-tility was still unexplained, making it difficult to draw conclusions about how or even if the hormone treatments worked. The results were good enough, however, to warrant further study.

Only one serious problem had developed over the course of the experiment: The women taking the hormones were often convinced that they were pregnant because the hormones produced many of the same symptoms as pregnancy: the women became nauseated; their breasts grew larger and more tender; and they stopped menstruating. Rock kept fielding phone calls from his excited patients saying that they thought they were finally pregnant. The women were desperate to have children. Most of them had been trying for years. "They were assured conception could not occur during the treatment," Rock said, but that didn't matter. The women kept calling and he kept telling them, sadly, that it was only the progesterone creating the illusion of pregnancy.

When he learned of Rock's work, Pincus was pleased but not sur-prised that the progesterone was having a contraceptive effect. The important thing to Pincus was the plain fact that Rock's patients were not dying. Here was proof, it seemed to him, that it was safe to give progesterone to women.

Rock told Pincus that while he was encouraged by his work with progesterone, he had one big problem: Patients receiving the hormone believed they were pregnant no matter how much he assured them it was not possible, and they were crushed when the truth finally became clear to them. He wished he could do something to ease their anguish.

Pincus proposed an elegant solution—and one that would have enormous consequences for his own work and for the future of women around the world.

A woman's menstrual cycle is usually twenty-eight days. Each month, estrogen and then a mixture of estrogen and progesterone flood the uterus, making the lining thicker in case it should receive a fertilized egg. Then, if no fertilized egg is implanted in the lining, hormone levels drop and the lining is washed away in a menstrual bleed. That's what Rock's patients were missing—the menstrual bleed that signaled to them that their bodies were functioning as normal and that they were not pregnant.

The question then was how to keep the women from ovulating while permitting monthly menstruation. The simplest solution was to have the women stop taking progesterone pills for five days each month. Their hormone levels would return to normal and they would bleed. It made sense to both men because it seemed so natural. Pincus, however, may have had another motive.

One of the companies supplying hormones for his progesterone experiments was G. D. Searle. Though the drug maker had been losing patience with Pincus a year earlier, company officials had never entirely given up on the brilliant but unpredictable scientist. Searle officials continued to believe Pincus might come up with something useful, and it was a lot less expensive for the drug company to write grants for the Worcester Foundation than it would have been to hire its own researchers. For those reasons, despite Pincus's failures, Searle agreed to pay the Foundation $62,400 for a twelve-month period beginning in June 1953. In addition, Pincus would receive shares of Searle stock, starting with nineteen shares valued at $921.50.

Pincus was more than a lab ace for hire. He frequently wrote to officials at Searle with ideas for new products, as he did when he heard about scientists experimenting with a liquid that was injected under the skin where it would solidify into a pellet. The pellet would then release hormones over a period of three to five weeks. The scientists in question were developing this liquid pellet to neuter chickens, which made them fatter and more tender. Pincus wondered if there might be human applications worthy of consideration and study. He also approached Searle about funding a cure for hair loss in men, a project he referred to as "Operation Baldness." Pincus understood that testosterone triggered hair loss in men, and he thought that injecting men with sex hormones to counter the effects of testosterone might prevent or perhaps even reverse hair loss. Searle officials expressed "grave doubts," but ultimately they gave him a green light.

In some of his letters to Searle, Pincus didn't mention progesterone or birth control. He had a business and lab to run and other ideas to pursue. But he also knew that helping men to regrow hair, while potentially enriching, was not going to earn him the respect he felt he deserved in the scientific community. Birth control offered the best chance for that.

Officials at Searle had not committed to a new contraceptive, and they asked Pincus not to publicize their involvement in his project. They also told Pincus that they would not have anything to do with a birth-control pill that interfered with the menstrual cycle. To change a woman's cycle, one Searle official said, perhaps sensitive to the concerns of the company's Catholic customers, would be "going against Nature."

It was a peculiar place to draw a line in the sand. Menstruation, which can be debilitating for some women, is only necessary when women are interested in getting pregnant. Wouldn't reversing baldness also go against nature? Doesn't penicillin go against nature by fighting infection? Even more to the point, doesn't progesterone go against nature by shutting down ovulation?

It didn't matter. If Rock's patients wanted to have their periods and Searle wanted women to have them, too, Pincus would see what he could do.

ELEVEN

What Makes a Rooster Crow?

THE HISTORY OF endocrinology might be said to have begun on February 8, 1849, at a meeting of the Royal Scientific Society in Göttingen, Germany. It was there that the scientist Arnold Berthold told his assembled colleagues about an unusual experiment in which he had castrated six young roosters. Without their testes, Berthold found, the roosters quit crowing. They also gave up trying to mate with females and stopped fighting with males. They seemed to lose the very essence of what made them roosters.

After observing these changes, Berthold replanted the testes in some of the birds. Suddenly, their combs and wattles grew back. They started crowing, fighting, and trying to mate. The results suggested to Berthold that within the testes of the rooster there must be some substance being released into the blood that affected behavior and body functions.

Forty years later, in 1889, an eccentric seventy-two-year-old French scientist named Charles-Édouard Brown-Séquard had much the same notion. But Brown-Séquard went about exploring it differently. He extracted fluid from the testicles of freshly killed dogs and guinea pigs and injected it into his own body. Almost instantly, he reported, he felt like a new man. The scientist reported that the injections invigorated him, sharpened his intellect, relieved his constipation, and even increased the strength of his urine flow. Scientists today believe Brown-Séquard was the beneficiary of a placebo effect, but his experiment nevertheless gained wide attention and inspired other researchers to explore the secretions of internal organs.

By 1905, scientists were learning about the body's endocrine glands, which include the pituitary gland, thyroid gland, pineal gland, parathyroid glands, thymus, pancreas, testicles, ovaries, and adrenal glands. The glands are small capsules of tissue. The biggest of the bunch, the pancreas, weighs less than three ounces. The pineal gland is no bigger than a grain of rice. Endocrine glands produce chemicals that serve as messengers to all the body's cells, controlling how we digest food, how fast our hearts beat, how we fight illness, when we feel depressed, when we reach puberty, and how we reproduce. The glands do all of this miraculous work by generating chemical compounds called *hormones*, from the Greek word meaning "to incite activity." Over the course of a woman's life, she produces barely one-fifth of an ounce of the hormones known as progesterone and estrogen, but that's enough to guide her reproductive system—and keep the human race in business.

It was the scientist Ernest Henry Starling who speculated in *Lancet* in 1905 that humans might be able to tinker with hormone levels to "acquire an absolute control over the workings of the human body." Soon afterward, scientists (and con men) began transplanting monkey and other testes into men, promising all sorts of rejuvenating effects. In Kansas in the 1920s, Dr. John R. Brinkley ran radio ads boasting that he could make even old men virile again by transplanting goat gonads into their bodies. The procedure had one surefire effect: it made Brinkley a rich man, albeit temporarily. By the 1920s, respectable physiologists and chemists had learned how to extract hormones from animal glands, as well as from bile and urine. They tried using the extracted hormones to compensate for the malfunction of human glands. In diabetics, for example, the pancreas fails to produce enough of the hormone insulin; the body needs insulin to regulate the amount of glucose in the blood. Diabetes almost always led to death until scientists in the 1920s discovered that injections of insulin could effectively treat the disease. After that, a great race was on to discover how other hormones might be harnessed to improve human health.

Three decades would pass before Gregory Pincus considered using hormones to control fertility. But he wasn't the first. In 1921, an Austrian gynecologist removed the ovaries of pregnant rabbits and guinea pigs and transplanted them into animals that were not pregnant. The animals were rendered temporarily sterile. The gynecologist conducting the experiment noted at the time that the same trick might work for women, although he didn't try it. In 1937, three scientists at the University of Pennsylvania tested progesterone for contraceptive benefits and found that the hormone put a stop to ovulation in rabbits. But, once again, no one dared try it on humans.

Contraceptive research was too controversial. In many instances it was illegal. As a result, there was little money in it. World War II did not help. It was difficult to make a case for the importance of family planning when mothers were losing sons and every able-bodied scientist was supposed to do the same thing as every able-bodied young man: join the fight for democracy. Scientists did just that. Physicists worked on the atomic bomb. Chemists worked on high-octane fuels that would help fighter jets fly faster. Biologists worked on hormone treatments to reduce stress in combat troops.

In the years after the war, many world leaders became concerned about overpopulation. The human race was growing too rapidly, and the world would run out of resources if something wasn't done about it. Poverty and hunger would spread rapidly. In the struggle for natural resources, countries could return to war.

In 1948, before going to work as the director of Planned Parenthood, ecologist William Vogt published *The Road to Survival*, warning that civilization might collapse if something were not done to control the population explosion. "In areas like Puerto Rico, where three-quarters of the houses lack running water, current contraceptive techniques cannot possibly be effective," Vogt wrote.

> *Hindus, with their $19 annual income, cannot possibly afford contraceptive devices. A cheaper, dependable method that can be easily used by women is indispensable. If the United States had spent two billion dollars developing such a contraceptive, instead of the atomic bomb, it would have contributed far more to our national security while, at the same time, it promoted a rising standard for the entire world.*

World War II took some of the stigma out of birth control in another way: The U.S. Army had spent millions of dollars supplying soldiers with latex condoms to reduce the spread of venereal disease, and American soldiers were often encouraged to think of their time in Europe as a great erotic adventure, stirring up what the historian Mary Louise Roberts called a "tsunami of male lust." Many of them returned to the United States eager to continue the fun, whether with wives, girlfriends, prostitutes, or some combination of the three. By the 1950s, Americans were spending about two hundred million dollars a year on contraceptives, mostly on condoms. The vast majority of doctors approved of birth control for the good of families, although many of them were afraid to say so publicly.

On the surface, the early 1950s seemed like a period of calm in America. Kids wore Davy Crockett hats in imitation of the TV actor Fess Parker. Men wore Bermuda shorts and drank highballs (often mixed for them by their aproned wives). Pop music was still slow and sweet, although young musicians such as Elvis Presley were beginning to experiment with sexy dance moves and a throbbing rhythm-

and-blues beat. Inflation was low. Politicians were actually believed to be telling the truth. New interstates allowed drivers to move faster than ever. Democracy seemed secure, which allowed world leaders the luxury of turning to long-range concerns such as overpopulation. But beneath that placid landscape, seeds of rebellion had been planted. Men were returning from the battlefields and adjusting to the fact they were no longer gallant warriors and makers of history; they were working stiffs with lawns to mow and gutters to clean. Marriage rates exploded. And women, with their husbands home from overseas but now at work all day, sometimes found themselves bored with household chores. World War II had shown them they could do more than tend house, but now there were fewer jobs available to women. Were they supposed to stay home and iron shirts, dust the blinds, and prepare dinner? Was there nothing else for them to do? Their boredom and frustration would eventually fuel a fight for women's liberation, but for now it contributed to a very different social movement. In 1950, 3.6 million babies were born in America, compared to 2.6 million a decade earlier. The median age for marriage in the 1950s was 20.1, and the median age for a woman to have her first child was 21.4. Birth rates were rising for every racial, ethnic, and religious group in the country.

Although no one had yet named it, the Baby Boom was beginning.

Sex was bubbling to the surface of American life. It was becoming more casual, not to mention more profitable. In 1948, when the Popular Library reissued its 1925 bestseller *The Private Life of Helen of Troy*, the cover image of Helen showed her clad in a sheer dress, her nipples erect, with a Trojan horse seemingly aimed at her pelvis. "HER LUST CAUSED THE TROJAN WAR," read the banner headline atop the book. Pulp paperbacks became huge sellers, with even classic books such as *All Quiet on the Western Front* repackaged with images of half-naked women to suggest they were really stories of yearning and sexual perversion. *Confidential*, the scandal magazine, told its salivating readers that Frank Sinatra ate Wheaties between sex acts, Errol Flynn had a two-way mirror in his bedroom, and Liberace liked boys. Even comic books turned kinky, at least according to the best-selling book called *Seduction of the Innocent*, published in 1954, which claimed that Batman and Robin promoted homosexuality, Wonder Woman encouraged women to become lesbians, and Catwoman was a whip-wielding dominatrix.

"FILTH ON THE NEWSSTANDS," proclaimed a headline in *Reader's Digest* in 1952. But filth sold. The rules of dating were changing, too, allowing men and women to spend more time in each other's company. There was a growing sense that women were not the passionless creatures that Victorian-era advice books had made them out to be. That image began to change in the 1920s, when flappers wore short skirts and smoked and drank alcohol in public, and it changed even more dramatically during World War II, when women took up jobs formerly held by men and acquired some of the power and confidence that came from earning their own money. After the war, America seemed to enter a socially conservative era with traditional gender roles back in place but women did not settle easily into their old, subjugated roles. Romance in the 1950s became a sexual negotiation, even a competition. A casual date might end in necking (kissing), but petting (deep kissing and touching), heavy petting (deep kissing with touching below the waist), or petting under the clothes was generally reserved for steady dates. Intercourse was supposed to be reserved for married couples, but those rules were changing too, and for the most part women set the limits. They were the ones risking pregnancy. They were the ones whose reputations would be damaged if they became known for sleeping around. Inevitably, this led to tension.

On college campuses, panty raids became one sign of the growing sexual conflict. At the University of Wisconsin, five thousand students charged women's dorms, bugles blaring, in an attempt to steal bras and panties. In Missouri the governor was forced to call in the

National Guard to stop a mob of two thousand men who battered down doors and broke windows to gain access to women's dormitories. "All animals play around," said Alfred Kinsey, laughing off the panty raids. But university and government officials took them as a serious threat to authority and a brazen expression of sexuality. Students were beginning to challenge authority and question the rules of sexual behavior.

It did not yet qualify as a revolution, but young people were seeking more independence, getting more involved in politics, and asking why they should adhere to the same moral standards as their parents. Instead of yearning for more responsibility as they grew up, they yearned for more freedom.

TWELVE

A Test in Disguise

PINCUS WAS PART jazz musician, part entrepreneur, part scientific genius, making things up as he went along. He had a little bit of money to play with from Planned Parenthood. He had progesterone, which worked relatively well on rabbits and rats. And he had John Rock, who had used progesterone on some of his patients. Rock's patients were infertile women, so his work proved only so much. But it was a start.

In a progress report to Planned Parenthood dated January 23, 1953, Pincus made no mention of testing hormones on women. The "most useful method of contraception," he wrote, would be a pill that inhibited ovulation, fertilization, and implantation of the egg. It would have to be non-toxic and completely reversible, he continued. He asked for $3,600 a year for two more years of animal testing. For Pincus, it was a remarkably modest request, perhaps suggesting that he did not have high hopes for a dramatic outcome.

Paul Henshaw, the director of research for Planned Parenthood, noticed a few omissions in Pincus's progress report and responded with several pointed questions. Were there any signs of side effects in the lab animals? he asked. Was Pincus thinking about human trials any time soon? Finally, there was the matter of patents. If contraceptive research funded by Planned Parenthood led to "patentable discoveries," Henshaw wrote, patent rights should be assigned to the Dickinson Research Memorial fund, which was part of Planned Parenthood. Henshaw, unlike many of the others who worked at Planned Parenthood, was a highly qualified scientist—a biologist who had worked on the Manhattan Project and had visions of creating a full-blown institute for the study of fertility under the Planned Parenthood umbrella—although he was already getting frustrated. Like Sanger, he found Planned Parenthood's leaders unwilling "to accept or tolerate the forward thinking required by research." Henshaw also recognized that Pincus's work might be profitable as well as patentable, and he went on record saying the organization wanted a share of any money generated by Pincus's research. "I should like to know whether you would agree to such a provision," he wrote.

No lawyers were involved in the discussion, only scientists.

Pincus did agree, with some caveats. He noted first of all that the contraceptives being tested might not be patentable because the chemical compounds under investigation came from various drug companies that had probably already submitted patent applications. But if research did produce a patentable birth-control pill, he continued, "a sharing of the patent rights seems to be the just and equitable procedure," given that the Worcester Foundation had spent some of its own money and money from other grants on the same research. "I admit this is a knotty question," he said.

In reply to Henshaw's other questions, Pincus said he had seen no evidence of side effects in animals receiving progesterone. Finally, he said he was indeed ready to begin human trials, but he was waiting to hear if Planned Parenthood intended to fund his continued research before making plans.

Two weeks later, Henshaw approved Pincus's request for two years of funding at $3,600 a year. When, he asked, can the human testing begin?

Pincus thanked Henshaw for the money and noted that he was beginning to develop a plan for testing the birth-control pill on large numbers of humans. "Most of this work is bound to be rather slow and laborious so that I do not anticipate any definitive results for at least a year," he wrote. Unless, of course, he added, Planned Parenthood could provide additional money. In that case, "one might get things going a little faster. . . ."

※

Pincus was pleased with the early tests on rabbits and rats and equally pleased, if not more so, with John Rock's work, which showed that progesterone caused no serious harm to women. Rock wasn't testing the hormone directly as a method of birth control, but that was a minor concern to Pincus. Whether the women were fertile or not didn't matter. He needed warm bodies. He needed women willing not only to take an experimental drug but also to submit to daily tests of their body temperature, daily vaginal smears, urine tests every forty-eight hours, and occasional endometrial biopsies, which involved doctors taking small samples of tissue from the lining of their uteruses. Rock's patients were motivated to do so because they believed that in the long run Rock would help them get pregnant. Without such an incentive, it was unlikely that Pincus would have been able to round up volunteers. Also, by billing the experiment as part of Rock's fertility work, Pincus could reasonably claim that he was not engaged in the administration of birth control. Had he been transparent in his objective, he and Rock both would have been in violation of the Massachusetts law banning all forms of contraception, and they would have been subject to prison sentences of five years and fines up to one thousand dollars.

In his application for additional Planned Parenthood grant money, Pincus wrote that he intended to conduct the tests through "two to three menstrual cycles in thirty to forty women." Meanwhile, lab tests on animals would continue.

Here was another one of Pincus's great improvisational bursts—a scheme so bizarre it could have come from Hollywood. He intended to test a birth-control formula by calling it a fertility treatment. If he had still been on the faculty at Harvard, or even if he had still been operating in affiliation with Clark University, he never would have gotten away with it. A university department head, fearing legal trouble or bad publicity, might have forbidden him from giving contraceptives to women. But Pincus was a free agent and unafraid to take risks. When Henshaw asked if Pincus was running the risk of breaking the law by giving birth control to women in Massachusetts, Pincus gave a stern response: "The fundamental facts may be established in Worcester or Timbuckto [*sic*]," he wrote. The tests were not for birth control, he continued. "They are concerned with the quite specific effects of the compounds with which we have been working, and a study of the biology of these effects is not against any law existing in Massachusetts." Testing on a large number of women in Massachusetts might attract unwanted attention, he admitted, but that was a matter for another day. The first thing to do was to find out if progesterone worked. He concluded: "I should therefore like to learn from you just what you think is possible in view of your available resources."

Pincus believed that it was the scientist's duty to be aggressive. Too many of his colleagues, he complained, were satisfied writing for scientific journals when they should have been thinking about how their work could be used to effect change. "The call-for-action programs," he wrote, "have largely passed over the research laboratory."

Pincus saw himself as more than a research scientist now. He was an activist, a crusader, and a businessman. He was also a builder of coalitions, however unlikely. He was not the first to experiment with progesterone or even the first to propose that progesterone might work as a contraceptive in women. He was simply the first to make the necessary connections, bringing together gynecologists, drug companies, and biologists with shared interests. By now, Pincus wasn't looking for a "eureka" moment; he was looking for pieces of the puzzle that might fit the rough image he had in mind for a functioning birth-control pill. But instead of keeping the project to himself or giving up because he had hit another roadblock, he pushed on, using every tool and every ally at his command.

John Rock was already giving progesterone and estrogen to women to see if it would help them get pregnant. Pincus wasn't doing anything different—he was merely doing it for a different reason.

Was it dishonest? Most would say yes. But it did not violate any of the laws or medical standards of the day.

In the 1950s the United States had some of the most progressive laws in the world when it came to testing experimental drugs, but there was still no law on the books requiring doctors to inform patients they were being included in an experiment. Rock's patients were not exactly hoodwinked. They were told that the progesterone they were receiving would shut down their ovaries and make it impossible to get pregnant. They were told the treatment would simulate pregnancy and might cause nausea. And they were told, honestly, that Rock believed they would have a better chance of becoming pregnant when the experiment was complete.

There was only one small piece of information missing.

※

Beginning in 1953, Pincus and Rock enlisted twenty-seven of Rock's patients at the Free Hospital for a three-month trial. This study would be different from Rock's previous experiments with progesterone. This time, because Pincus wanted to be certain that the hormone was effective in halting ovulation, women who failed to ovulate regularly were not included. The women enlisted were still infertile, but Rock didn't know what was causing their infertility. Instead of the progesterone-estrogen mix that Rock had used in the past, the women in this experiment received only progesterone. And they received the tablets daily for only three weeks out of each month, stopping for a week to allow them to menstruate.

The tests were demanding. Among nurses and lab workers in Rock's office, the new round of testing was referred to as the Pincus Progesterone Project, or the PPP. Some of the staff joked that PPP stood for "pee, pee, pee," because so many urine specimens were tested. When enough bottles of urine were amassed, Rock would send a technician from Boston, and Pincus would send a technician from Shrewsbury. They would meet somewhere in the middle, with Pincus's man taking the urine and driving it back to Shrewsbury.

Pincus was so excited about this approach that he couldn't wait to enroll more women. In the spring of 1953, as Rock's trials were beginning, he approached scientists and gynecologists in Israel, Japan, and Worcester, asking if they would enlist their patients in similar studies. He also recruited nurses at the Worcester State Hospital to participate. The nurses might have been hugely important to Pincus because, unlike Rock's patients, they were not being treated for infertility. Presumably they were fertile women, some of whom were using other forms of birth control. Unfortunately for Pincus, the nurses turned out to be lousy patients. Most of them dropped out.

In Worcester, a forty-seven-year-old gynecologist named Henry Kirkendall agreed to help Pincus by enrolling some of his patients in the study. Kirkendall was a member of the staff at St. Vincent Hospital and the senior obstetrician at Memorial Hospital, both of which were in Worcester. If you were born in or around Worcester in the 1940s or early 1950s, there was a strong chance it was Dr. Kirkendall who delivered you. Like John Rock, Kirkendall was a staunch Catholic and, also like Rock, his work with women had made him sympathetic to the cause of family planning. Pincus met with Kirkendall and asked the gynecologist if he could find thirty women willing to participate in a study similar to Rock's. The women would need to take their own temperatures every day and record the findings. They were also expected to take daily vaginal smears and collect their own urine for hormone analysis or else visit the doctor's office to have it done by a nurse. The progesterone dosage would be extremely high—between 250 and 300 milligrams a day. The women were not paid for their participation, nor were they informed that the results might lead to the invention of a new form of birth control. Most of them were doing it simply because a trusted doctor had asked them to.

The tests began in June. All through the summer, Dr. Kirkendall would load the trunk of his baby-blue Pontiac convertible with vials of urine and slides containing vaginal smears and drive them over to the Worcester Foundaton in Shrewsbury for testing, or he would ask his son to drive them there.

In the first year of trials, Pincus, Rock, and Kirkendall enrolled sixty women in their study. That alone was something of an accomplishment, given that they'd been forced to go about their work surreptitiously. But half of the women enrolled dropped out along the way, either because the test procedure was too demanding or because the side effects were too disturbing. And while Rock was pleased with the results—four of the thirty infertile women completing the study got pregnant on the so-called rebound—Pincus was disappointed: About 15 percent of the women showed signs of ovulation while taking progesterone, significantly worse results than he'd seen with rabbits and rats. To complicate matters, the results were the same for women whether the dosages of progesterone were two hundred milligrams or four hundred.

Until that moment, everything had been happening quickly and encouragingly. But a contraceptive with an 85 percent success rate was no good to anyone.

Suddenly, Pincus had reason to doubt his own elegant solution.

THIRTEEN

Cabeza de Negro

GREGORY PINCUS DIDN'T know that the missing piece of his scientific puzzle had been discovered a decade earlier. No one knew it yet.

In 1942, weeks after the Japanese bombed Pearl Harbor, an American chemist named Russell Marker boarded a train for Mexico in search of an enormous root plant the locals called *cabeza de negro*, because from a distance the round part of the plant that stuck out of the ground looked like the top of a black man's head. The American embassy had been advising citizens to stay out of Mexico, Marker recalled, "because no one knew whether we were going to get into the war or whether Mexico would be in on it and on our side or not." Marker's employer, Pennsylvania State College (now known as Pennsylvania State University), also urged him to postpone his research mission, but Marker refused. He traveled to Mexico by train, carrying with him a map from a botany book that indicated *cabeza de negro* could be found in the Veracruz province of the country where the road between Orizaba and Córdoba crosses a river gorge.

In Mexico City, Marker, who spoke no Spanish, hired a local botanist and the botanist's girlfriend to accompany him by truck to Veracruz. But after three days on the road, the botanist and his girlfriend became frightened, saying there was too much animosity toward Americans in the area. It was unsafe to go on. Marker drove back to Mexico City with them and then boarded a rickety old bus bound for Veracruz, sitting next to a woman carrying live chickens. When he arrived in Orizaba, he switched to another bus, this time headed for Córdoba. Spotting a small stream between the two towns, he asked the bus driver to stop and let him off. Not far from the bus stop Marker found a small grocery store, went in, and asked the owner in English if he knew where to find *cabeza de negro.* The storekeeper told him to come back the following day—*mañana*, he said. That much Marker could understand. When he returned in the morning, the storekeeper presented him with two plants, which Marker put in bags and tied to the roof of the bus that carried him back to Mexico City.

※

Marker was a maverick. He was prematurely bald, powerfully built, and chronically impatient. At twenty-three, just shy of earning his doctoral degree in chemistry, he'd quit because the University of Maryland had required him to take classes in which he had already earned a master's degree. He couldn't bear the thought of wasting his time. Without a doctorate, Marker bounced around the world of science, from the Ethyl Corporation (where he helped develop the octane-rating system for gasoline still used to this day) to the Rockefeller Institute. While at the Rockefeller Institute, he became interested in hormones. Marker observed that scientists were doing interesting work on hormones, but most hormones remained scarce and prohibitively expensive. The high prices stymied research. Progesterone at that time was so expensive that research scientists couldn't afford to experiment with it at all. Even if the scientists could afford small amounts of the drug and even if their research led to fantastic breakthroughs, patients in need of the newly developed drugs would never be able to afford them. But Marker had a hunch that progesterone and other hormones could be made much more cheaply from vegetables. He mentioned this to his boss at the Rockefeller Institute, Dr. P. A. Levene, but Levene told him it had already been tried and hadn't worked. Marker refused to accept such an abrupt dismissal. "I said it was a practical thing to work on and that if I couldn't work on them at the Institute, then I would find a place I could work on them," he recalled. He quit and accepted a position with Pennsylvania State College, taking a massive pay cut from $4,400 to $1,800 a year.

For a chemist, sex hormones such as progesterone and estrogen are relatively easy to manipulate, which is why they intrigued Marker. Sex hormones are steroids, organic molecules that share a common structure of carbon and hydrogen atoms arranged in four fused rings. Thousands of hormonal compounds have the same structure. The structure of progesterone, for example, is similar to that of testosterone. Their simplicity makes modification easy. With a few chemical steps, male hormones can be converted to female hormones and female hormones into male. New discoveries were coming rapidly as Marker began to dabble in this burgeoning field. At Penn State, he managed to make thirty-five grams of progesterone from the urine of a pregnant woman. At the time, doctors were beginning to treat women who suffered frequent miscarriages by giving them small doses of progesterone. The president of Parke-Davis, a drug company that helped finance Marker's work, was thrilled, saying the company could sell Marker's progesterone at a thousand dollars a gram. Marker soon figured out a way to do even better by making progesterone from bull urine. Meanwhile, other scientists found they could make sex hormones by modifying the chemical structure of cholesterol, although this method still proved expensive and time consuming. By 1940, Marker made a huge breakthrough and confirmed the hunch he'd had while working for the Rockefeller Institute, developing a five-step chemical process to produce progesterone from a compound found in the sarsaparilla root. His search for a plant that would yield even more of the key compound led him to Mexico and the *cabeza de negro*.

Soon scientists all over the world began manufacturing progestins (the term for synthetic versions of the natural hormone) and trying to improve on Marker's work. Marker helped found a new company, Syntex, which quickly became one of the world's leading suppliers of progesterone. But true to character, Marker soon became frustrated, saying he had never received his share of the profits generated from his work. When he quit, he destroyed all of his papers. His replacement, a Hungarian Jewish chemist named George Rosenkranz who had fled Europe during the war, struggled to replicate Marker's work. Marker had been secretive about his methods and never even bothered to label the chemicals in his lab, relying on sight and smell to tell one from another. Within five years, though, Rosenkrantz not only matched Marker's work but also worked out the synthetic production of androgens and estrogens. In 1949, Rosenkrantz recruited a young Austrian-born American named Carl Djerassi, who had left the University of Wisconsin to work in Mexico because he had heard that Russell Marker and others were doing revolutionary work there. Djerassi was determined to improve the company's progestins.

For reasons not completely understood, synthetic progesterone was not particularly effective when taken orally. Injections were painful, and even those required larger doses than most other sex hormones. Djerassi set out to create a more potent progesterone that would work when taken orally. He remembered a paper he'd read in college describing how a chemist named Max Ehrenstein had removed a carbon atom from a molecule and replaced it with a hydrogen atom. To Djerassi, it was as if Ehrenstein "had transformed a very elaborate mansion . . . into a funky little vacation house." It was a highly inefficient operation, but it gave him ideas. His modifications produced a new compound four to eight times more potent than the old ones. Perhaps best of all, his new compound was also able to survive absorption in the digestive tract, which meant it could be taken orally. Djerassi thought the new compound might be effective in helping women with menstrual disorders. He didn't know that another young chemist—Frank Colton of G. D. Searle & Co.—had also been inspired by Ehrenstein to develop virtually the same thing. And neither Djerassi nor Colton knew that Gregory Pincus was searching for exactly such a compound.

"Not in our wildest dreams," Djerassi said, "did we imagine that this substance" would one day become the active ingredient in a birth-control pill.

FOURTEEN

The Road to Shrewsbury

MARGARET SANGER AND Katharine McCormick were counting on Pincus more than ever. Other researchers were exploring new forms of contraception, but most of the work involved sloppy stuff such as foams and jellies that did nothing to promote sexual spontaneity or pleasure. Sanger wanted a more precise instrument, something wholly new and groundbreaking. Yet so far neither McCormick nor Planned Parenthood had come up with enough money to capture Pincus's full attention and ensure that he would see the research project through to the finish.

In May 1953, as Pincus and Rock were launching their first round of tests on humans, Sanger suggested that she and McCormick pay a visit to Shrewsbury to meet Pincus and see if he was someone McCormick might wish to support. McCormick wrote back that she would be delighted to make the trip. Her only request was that the journey be planned for a weekday, because, as she wrote, "I do not like to plan for any motoring on Sundays for it simply means waiting in a long line of cars, as the crowds are so great on that day."

The meeting was set for June 8—a hot and humid Monday. McCormick's chauffeur-driven Rolls-Royce glided through the rolling hills of central Massachusetts, past the sturdy working-class homes of

Shrewsbury, and onto the macadam driveway of the Worcester Foundation.

The women were not put off by the utilitarian, cheaply furnished offices; the poor ventilation in the animal rooms; or the jerry-rigged appearance of the laboratories. If anything, they seemed captivated by the place's scrappy charm, and they were especially enthusiastic about Pincus's plan for testing John Rock's patients. At one point, Chang heard McCormick whisper in Sanger's ear: "This is the place."

As the tour concluded that afternoon, McCormick asked Pincus how much money he needed. He had already negotiated a grant of $17,500 from Planned Parenthood to cover the first year of clinical trials, and Planned Parenthood had already turned to McCormick for help in paying for the work. McCormick had agreed to pay half of the $17,500.

Now, as she strolled the grounds of the Worcester Foundation, McCormick asked for the big picture. How much would it take to fund the entire research program? How much would it take to get her pill?

Pincus answered: $125,000.

McCormick nodded and thanked Pincus for his time. She and Sanger got back in the car and returned to Boston.

The next day, McCormick phoned Pincus to say she would write a check for $10,000, with more to come. That night, as Pincus might have been relaxing at home, celebrating with a cocktail, and telling Lizzie that at last he would have enough money to commit real effort to the search for an oral contraceptive, the skies turned a sickish purple as a terrifying tornado tore across Worcester and Shrewsbury. The tornado killed nearly one hundred people, injured more than eight hundred, and left thousands homeless. Entire houses were lifted from their foundations and relocated. Couches and refrigerators were sucked through windows. Sheets of sidewalk flew through the air as if weightless. A brand new tool factory was reduced to a jumble of brick and metal. At the Worcester Foundation, a roof was destroyed on one building and a small amount of lab equipment suffered damage when windows shattered. But no one was hurt.

For Pincus, it was a week to remember.

FIFTEEN

"Weary & Depressed"

KATHARINE McCORMICK WAS always impatient. Once, she complained about Christmas—"the long stretch of holidays irritates me"—because no work gets done.

Now, in the fall of 1953, she was even more impatient than usual. "I haven't heard a word about the Pincus work," she complained to Sanger in a letter dated September 28. Only three months had gone by since her meeting with Sanger and Pincus at the Worcester Foundation in Shrewsbury, but already she feared the worst. "I do hope they have not run into difficulties," she wrote. "There has now been time enough . . . to get some sort of an idea as to how progesterone could function. I am most anxious to hear what they think of it and of the scope of the tests in action."

A week later, Sanger sent word of a troubling rumor: Planned Parenthood did not intend to fund Pincus's research beyond January 1954. Sanger said she had been unable to confirm the rumor, but she was angry and concerned. Pincus's strongest advocate within the Planned Parenthood administration, Paul Henshaw, had been fired recently after a power struggle with William Vogt. Sanger complained to McCormick that the organization's leaders were paying no attention to "the greatest need of the whole P.P.F. movement . . . that of a simple, cheap, contraceptive." And if all that weren't enough, Sanger got news that Planned Parenthood had named John Rock—"the ardent Roman Catholic," as she called him—chairman of its research committee. "You will recall what a very charming person John Rock is," she wrote, "but that does not mean his interest in contraception is sufficient for him to become a Chairman of a Research Committee, whose object should be the discovery or the development of a simple contraceptive."

Sanger urged McCormick, who had recently moved from California to Boston, to pay Pincus another visit and see what she could learn. "His work, it seems to me," Sanger wrote, "is the nearest to reality and it would be a shocking and devastating happening . . . if he should not be able to bring it to a final conclusion."

Pincus could see the same issue Sanger did, that Planned Parenthood was not fully committed and might not ever fully commit. As a result, he was hedging his bets.

The scientist had earned his way back into the good graces of G. D. Searle & Co. In launching his quest for a birth-control pill, he had made it a point to use Searle to supply many of the progestin compounds he was testing, trying one after another to see which worked best in much the same way a chef might try subtly different spices in a new dish. Carl Djerassi at Syntex and Frank Colton, Searle's scientist, had both discovered how to make synthetic forms of progesterone, but each compound had a slightly different chemical structure and each produced slightly different results. Some worked well by injection but not orally. Some required higher doses than others to be effective. Pincus and Chang worked their way through thousands of laboratory rats in the exploration process, hoping as they worked that the results in humans would track with those in rats. Of the three progestins Pincus liked best—known by their chemical names norethynodrel, norethandrolone, and norethindrone—two were made by Searle. "The Searle Company seems to have a bit of luck these days since we have tested a large number of other compounds," he wrote to Al Raymond, the man who had all but declared Pincus worthless two years prior. Years later, one scientist looking at the data from Pincus's early tests said he was surprised that Pincus had chosen Searle's compound because the one made by Syntex had performed better. If Pincus were biased toward Searle, he had good reason. He was not confining his activity to progesterone research. He couldn't be sure the funding Sanger and McCormick provided would continue. Even if the funding did continue, the results were difficult to predict. And even if he got good results with Rock's patients, he had no idea what would happen when it came time to test the new drug on thousands of women at a time. Was such a thing even possible in a country that still restricted access to birth control? Would foreign countries allow American scientists to come in and experiment on their citizens? The entire project could crash to a halt at any moment.

In the meantime, Searle was funding the Worcester Foundation at a rate of about $5,600 a month, making it the Foundation's biggest private backer by far and accounting for about 8 percent of the Foundation's total income.

Searle executives were interested in Pincus's progesterone research, but they were interested in his other activities as well. In the summer of 1953, Pincus, clearly optimistic about his standing with the big drug company, and perhaps optimistic as well about his birth-control pill or his baldness cure, purchased nineteen shares of Searle stock at $48.50 per share. He asked the company to deduct the cost of the stock from his pay at a rate of $18.43 a week.

At the time, Searle was paying about one-third of Pincus's $15,000 annual salary when the median family income in the United States was $5,000, Mickey Mantle earned $17,500 to play centerfield for the New York Yankees, and President Eisenhower received a salary of $100,000.

※

Pincus may have been hedging his bets, but Sanger and McCormick were not. They were increasingly desperate to see him succeed.

Sanger was "weary & depressed," as she wrote to her friend Juliet

Barrett Rublee, with little time for family and friends. "I feel pepless—no energy—no desire to do anything," she continued.

Despite decades of work toward the cause of family planning, her abrasive nature had left her without strong support from the very organization she had helped to found. She was often forced to make personal fundraising appeals and to dip into her own bank account to fund projects that Planned Parenthood would not fully cover. "God knows how far I can go spending for Bc [birth control] & Conferences," she wrote to the same friend.

Sanger was getting sicker by the day as heart disease sapped her strength. She had suffered her first attack in 1949, shortly after her seventieth birthday. Afterward, fatigue forced her to cancel appearances. In January 1952, when she made her television debut, her Catholic debate opponents scarcely let her get a word in, and Sanger looked like an aging prize fighter as she absorbed blow after blow without hitting back. "I am so discouraged at my own reflexes—& lack of 'come back,'" she wrote, "that I have almost decided not to do any public speaking ever again." She relied on Demerol, a highly addictive painkiller, to relieve severe chest pain. Her doctors urged her to retire, but she gave them a one-word response: "Preposterous!"

Friends noticed that she seemed more irritable than usual, perhaps a result of the mix of painkillers, sleep medicine, and the champagne to which she had become accustomed. "I doubt very much you are at all aware in your concentration on the work, of how you brushed your good friends off and made them feel completely unwanted," Dorothy Hamilton Brush, Cleveland-born socialite and Sanger's longtime friend, wrote to her in 1953.

When she couldn't sleep, Sanger read *The Second Sex*, the new book by Simone de Beauvoir, which would become one of the founding texts of feminism, weaving together history, biology, economics, and philosophy in an exploration of the power of sexuality. De Beauvoir's book attempted to explain why women had accepted a secondary role to men in society from the age of hunters and gatherers to World War II. She argued that men and women would both be better off if women won their emancipation, writing: "It is when the slavery of half of humanity is abolished and with it the whole hypocritical system it implies that the 'division' of humanity will reveal its authentic meaning and the human couple will discover its true form." It was no surprise that Sanger enjoyed the book, especially given de Beauvoir's descriptions of "the servitude of maternity," "woman's absurd fertility," and heterosexual love as "a mortal danger." De Beauvoir in her personal life did not hate men, but she did hate the institutions forced on women by a male-dominated society. Even conjugal love, she wrote, is "a complex mixture of attachment, resentment, hatred, rules, resignation, laziness and hypocrisy."

Sanger didn't say if she agreed on all counts, but wrote: "It is an amazing study of woman—it must have taken her most of a life time to study & write." The part she may have liked best was de Beauvoir's call to action. "What a curse to be a woman!" de Beauvoir wrote. But the greater curse, she went on to say, was to accept the curse without a struggle. It was every woman's duty to fight.

That had never been a problem for Sanger before, but now her time and energy were growing more limited. And rather than retire, as her doctor had urged, she took on greater responsibility, becoming the sole president of the International Planned Parenthood Federation, the organization working to spread birth-control work around the world. She remained frustrated that the Planned Parenthood Federation of America, focused primarily on birth-control education and clinical services, wasn't doing more to back Pincus's work. She was convinced that the scientist was on the brink of a major breakthrough. As she told the doctors and social workers attending the Fourth International Conference on Planned Parenthood in Stockholm in the summer of 1953, the time had come to "put all our energies into research for a simple cheap contraceptive, one which will perhaps immunize temporarily against pregnancy. I believe that will be the safest method and the one which in the long run will be the best." She went on, playing to the eugenics supporters in her audience, to say that the next priority should be to "do something definite about the breeding and multiplication of diseased families . . . mental defectives, morons, unhealthy, diseased people."

Sanger's comments were controversial, but not scandalous. In 1952, there were 1,401 officially reported sterilizations of so-called "mental defectives" in the United States. There were no doubt many more that went undocumented. Some states did not report the sterilization of prison inmates or the mentally ill, and other cases went unreported when individuals volunteered for sterilization. At the time of Sanger's speech, some advocates were calling for more compulsory sterilization; some called for a campaign to encourage more voluntary sterilization; and others opposed it for any reason other than individual matters of health. Sanger and others within the Planned Parenthood movement used the occasion of the 1953 conference to call for more research and education to bring about, as one speaker said, "a more general acceptance and intelligent use of sterilization . . . to benefit both the individual and the community."

Sanger and Planned Parenthood had always had an uneasy alliance with the eugenicists. In the 1920s, college professors taught courses on eugenics and students constructed state fair booths to teach visitors about "racial hygiene." The movement's leading spokesman, Charles B. Davenport, had opened the Eugenics Record Office in Cold Spring Harbor on Long Island, which became the movement's center. Some of the support for the movement came from social workers, health officials, doctors, and nurses who saw the tragic consequences of inherited disease, while others were compelled by racism and elitism to develop a biological program that would reduce the size of immigrant and racial groups they deemed less desirable. It was no great surprise that Sanger, who learned about eugenics from

Havelock Ellis, would find it attractive. "More children from the fit, less from the unfit—that is the chief issue of birth control," read a 1919 editorial in Sanger's *Birth Control Review*. She believed women should be empowered to control and limit their own reproduction. She also argued the government would not have to resort to welfare for the poor if society used the same efficient reproductive techniques as "modern stockbreeders" to improve the health of the populace. Parents, she said in a speech, should have to apply for the right to have children just as immigrants applied for visas.

The men who led the eugenics movement went further, arguing that the state should be empowered to control the reproduction of whole groups of people they deemed inferior or unworthy of the right to reproduce. They saw a danger not only in the growth of poor and unhealthy families but also in mixing of races and nationalities. In her ambition to get help from the eugenicists, Sanger may have overreached, or she may in fact have been a true believer. By the end of the 1930s, eugenics faded from public view, but Sanger had not given up on the movement's remaining supporters. In a 1950 letter to McCormick she wrote, "I believe that now, immediately there should be a national sterilization for certain dysgenic types of our population who are being encouraged to breed and would die out were the government not feeding them." Even after World War II, when the Nazis attempted to eradicate entire races and religions using sterilization and mass murder to accomplish their goals, Sanger held firm. "Parenthood," she said repeatedly, "should be considered a privilege, not a right." Sanger and McCormick were both elitists, to be sure, and they grew more elitist as they got older and wealthier, but there's little reason to believe either one of them was racist. Sanger never joined the eugenicists who argued that rich, educated, white people should be encouraged to have *more* children. Nor did she single out any race when she identified people who she felt ought to be having fewer children. She wanted women to have fewer children, or to have the best odds of having healthy children when they wanted them. Race never seemed to be the driving factor in her deliberations.

Regardless of her motives, Sanger's loyalty to the eugenicists presented a dilemma, because a birth-control pill was not really what the eugenicists wanted or needed. As some of the eugenicists were savvy enough to point out, a birth-control pill, no matter how inexpensive, would probably appeal most to well-educated and wealthy women. These were precisely the women that eugenicists wanted to see having *more* children, not fewer.

Sanger had begun her crusade as an advocate for the poor and disenfranchised, but in cozying up to the eugenicists she had effectively converted it, as the historian David M. Kennedy wrote, "from a radical program of social disruption to a conservative program of social control." By the 1950s Sanger seemed to recognize the problem of being so closely linked with the eugenicists, but it was too late. If she wasn't quite married to them, she'd been in bed with them so long that there was no way to call it off. Over the course of her long career she had done a great deal—perhaps more than anyone in the twentieth century—to change attitudes surrounding family, women, and sex, but most of the change had occurred among the middle and upper classes. Women with education and economic standing were more likely to stand up for themselves and discuss family planning with their husbands than women from the lower classes, as the sociologist Lee Rainwater wrote in 1965 after conducting more than four hundred interviews on the subject. The poor, meanwhile, were not much better off than they had been at the start of the century, when Sanger had been canvassing New York's Lower East Side and tenement women had been forced to rely so heavily on abortions to control the growth of their families.

In the end, it wasn't simply the eugenicists who had led Sanger away from her original goal of helping the poor; sex was a factor, too. If her goal had been simply to help the poor, she might have stuck to education. But she had wanted to liberate sex for women of every class. She had wanted sex to become a greater source of pleasure and personal fulfillment. She had wanted to see it deepen the bonds between men and women. She had wanted to curb the world's population growth. She had wanted it all. Indeed, she still wanted it all, and she still held out hope. It was all riding on Pincus's pill.

Sanger was seventy-four. The average lifespan for an American woman at the time was seventy-two years. Her age and poor health gave her good reason to contemplate her legacy and unfinished business. But in the fall of 1953 she had yet another reason to reflect on her triumphs, miscalculations, and unfinished work: A young journalist named Lawrence Lader had begun writing her biography. The attention thrilled her.

"You must always be in love," Sanger told the author, who was forty years her junior. "Life is meaningless unless you are in love." She insisted he stay close to her constantly as they worked, spending all day and all evening together and drinking terrific amounts of champagne. There was nothing physical between them, but to Lader it felt every bit like a courtship.

Lader was so charmed he sent Sanger chapters to review as he finished them. "I am not happy in past memories," she wrote him in October 1953 as she worked her way through his manuscript, striking through lines she didn't care for and inking notes in the margins. Lader, clearly smitten, let her make the changes. "In my revised opening I did use the . . . comparison with Joan of Arc and am realizing how smart Joan was to attribute her driving impulses to a thing as simple as voices," Lader wrote in a letter to Sanger. "But how does one put over the radiance, the inexhaustible flame of your own driving force?"

Lader's book, published by Doubleday in 1955, was almost entirely without criticism, and at times it was so admiring as to be embarrassing. Even so, as she read it and edited it, Sanger saw her life in summary. It was like reading one's own obituary, only longer and more intense. She compared Lader to "a dog with a bone, digging and digging into the past, and into the psychic experience of ones [*sic*] life that really could make you quite ill."

The book's final chapter made clear how this crusader's story would end: "Only one objective, one theme like an enormous orchestral crescendo, dominates Margaret Sanger's work today—the search for the birth-control 'pill,'" Lader wrote. "The discovery of the 'pill'—the climactic dream of her life—will undoubtedly prove one of the revolutionary events of the century. . . . Although Mrs. Sanger has played no technical role in the development of the 'pill,' she has been its prophet, its driving force." Lader went on to write that the pill might arrive within the next year or two, with testing to confirm its safety taking an additional three to five years. In other words, it was coming soon—soon enough, perhaps, that the fading warrior might yet be around to see it happen.

SIXTEEN

The Trouble with Women

PINCUS WAS STILL operating the Worcester Foundation as if it were a race car built to go quickly over short distances. In the fall of 1953, the Foundation appeared to be thriving, with forty-six grants providing a total of $622,000 in income. That was a lot of money for a relatively small, independent laboratory, but there was a catch. Pincus and his partner Hoagland were pouring almost every dollar possible into research while neglecting more mundane matters such as building maintenance and support staff.

The Foundation's business manager was exasperated. He'd been telling Pincus for years that the organization needed to devote 25 percent of income to overhead, but Pincus wouldn't listen. He kept overhead costs down to about 11 percent of the operating budget. The attic and basement were jammed with animals, as Pincus kept ordering more rabbits and rats regardless of whether he had any place to put them.

"Right now the directors want a new animal building," the business manager, Bruce Crawford, wrote in a 1953 memo to the finance committee, "and the need is really desperate, not just for the new work but to do the jobs on hand." Yet there was no money for construction, or anything else, because Pincus had refused year after year to set aside funds for future projects.

If Pincus had been trying to build an institution to last for generations, he might have considered launching a campaign to create an endowment, something typically considered essential at academic institutions. He might have hired a consulting company to advise the Foundation on establishing better business practices or hired an executive with a business background to run the organization as chief executive officer or chief financial officer. But Pincus did neither of those things. He was too focused on his research, on the next discovery. Hoagland, who spent more time than Pincus on operations, never tried to rein in his brilliant partner. So instead of developing a long-range business plan, the men went to Katharine McCormick and asked her for money to build an animal testing center, and she immediately agreed, pledging fifty thousand dollars. Problem solved.

Pincus wrote to thank McCormick, saying he was gratified to have the money but even more gratified by her strong interest in the work being done. As construction for the new building got underway, Pincus pursued more funding from big foundations, telling them that he would soon have increased capacity to do research. In one letter to the Josiah Macy Jr. Foundation, which had supported his work on *in vitro* fertilization at Harvard, he said he was working on a system that would allow women to freeze their eggs the same way men froze their sperm. Without mentioning birth control in particular, he bragged that whole new possibilities in the field were beginning to emerge. "It is my honest opinion that a science is being developed in this field, a science which is young, lusty and full of promise," he wrote.

Meanwhile, he continued to try different progestin compounds to see which worked best and continued to look for ways to test on women. It's difficult to tell from his notes and papers whether the birth-control work had become a great priority or whether he considered it a long shot, one among many on his agenda. In the final months of 1953, he was juggling more than a dozen projects, including studies on cholesterol, adrenalin, schizophrenia, and the metabolism of steroids. His notes and letters give no indication that the birth-control project was of any greater priority than the others. In fact, when he gave annual updates to the Foundation's trustees, he scarcely mentioned birth-control research, saying only that his current work included "new studies in reproduction control." Pincus understood the potential of a birth-control pill. He understood that it might not only change women's lives but also help tame the world's population growth. In spite of this understanding, he was not yet an evangelist for the cause; he was a scientist following good leads and chasing grant money, and if it had not been for the persistence and extraordinary beneficence of Katharine McCormick, the birth-control project might have gone nowhere.

More than ever, McCormick was taking charge. Sanger was in poor health, low on energy, and increasingly cantankerous. Her diaries indicate that she was spending much of her time seeing doctors, taking art classes, and helping to plan local festivals in Tucson. "I do not know how the Pincus project is progressing," she admitted in a letter to McCormick dated February 13, 1954. McCormick wrote back to say how disappointed she was by the performance of top officials at Planned Parenthood. William Vogt, the organization's director, had come to Boston and had failed to make time to visit Pincus in Shrewsbury. Vogt had met with McCormick but failed to ask for money and "merely said he hoped I was still interested in their work!" McCormick was mystified.

She followed up on the meeting in Boston by going to New York to visit Vogt at the Federation's headquarters. But while she had arranged the trip weeks in advance, upon arrival she found that no one from either the medical or research committee was on hand to speak to her. "As I became somewhat impatient," she wrote, "I could not help saying that I seemed to be the only person really interested in an oral contraceptive." In response, Vogt revealed one of the reasons he had been treating McCormick coolly. He thought she was wasting her money on the Worcester Foundation's animal house. Vogt had been gradually losing interest in scientific research, convinced that Planned Parenthood's money was better spent on education and clinical organization. McCormick was taken aback. She told Vogt that she believed strongly that the progesterone research offered the best chance for a birth-control pill but that it was also a distinct possibility that "an allied steroid," meaning something akin to progesterone, would prove even more effective. Since Pincus and his team in Worcester were the only ones testing the hormone's contraceptive powers and the early results were promising, she did not want the research to lapse. Even as human trials were getting underway, McCormick understood that continued animal testing would be essential to find the most effective chemical compound. She concluded that no one at Planned Parenthood was "really concerned over achieving an oral contraceptive and that I was mistaken originally in thinking they were." With Sanger at times distracted and at times incapacitated by illness, it was McCormick who urged the work forward. She was beholden to no one and was free to speak her mind. While Planned Parenthood officials and even Margaret Sanger tried to be careful to say that new forms of birth control were being sought only for married women, McCormick alone among the team of developers had the courage and independence to declare that all women, married or not, should have access.

Pincus and Hoagland were completely reliant on McCormick's direct support, but they cautioned her that it would be unwise to ignore Vogt and the others at Planned Parenthood. Though its leaders were often at odds with Sanger and sent mixed signals about Pincus's research, the organization remained a valuable ally, one that Pincus was not inclined to alienate.

Increasingly, McCormick dealt directly with Pincus. In November 1953, she made the drive to Shrewsbury to visit Pincus and ask when he and Rock would be ready to report the results of their first human trials. Pincus explained that they had been testing about seventy women since July—most of them patients of Rock's or Dr. Kirkendall's, plus a few nurses from the Worcester State Hospital—but nearly half the women had dropped out of the trials, complaining that the procedures were too rigorous or else that the treatments were making them nauseated.

"Human females are not as easy to investigate as rabbits in cages," McCormick wrote to Sanger in November 1953, summarizing her conversation with Pincus. "The latter can be intensively controlled all the time, whereas the human females leave town at unexpected times and so cannot be examined at a certain period; they also forget to take the medicine sometimes." And those weren't the only complicating factors. Unlike rabbits, women had to explain to their husbands why they were taking experimental drugs that might stop them from making babies. Unlike rabbits, women asked questions. And, unlike rabbits, women complained when they felt sick.

Testing a birth-control pill would be more difficult than testing other drugs. It was one thing to try a new medication for sick people. Sick people wanted to get better. Sick people were often under close medical supervision, either in clinics or hospitals. Sick people were willing to accept side effects and risk because the alternatives were worse. But these were healthy women—healthy *young* women—volunteering for the experiments. What if the pill made them sick? What if it made them permanently sterile? What if it caused deformities in their babies? What if it somehow altered a woman's hormones so that she could only give birth to girls?

To be certain that the medication was safe, Pincus told McCormick, he would need to test hundreds if not thousands more women. And to test more women he would need more staff—doctors, nurses, and clerks—as well as more examining rooms. Past experience told him McCormick might volunteer to fund the research, but there was one thing McCormick's money couldn't supply: patients. After nearly a year of work and a dropout rate of about 50 percent in their first round of testing, he and Rock had completed research on only about thirty women.

Psychiatric hospitals and gynecology practices would never be enough. But Pincus knew a place where women wouldn't ask many questions and wouldn't complain too much (or so he believed), a place where birth control was legal and widely accepted.

"Mrs. Pincus and I recently returned from a trip to Puerto Rico," he wrote McCormick on March 5, 1954. While there he had lectured at medical schools and before groups of doctors. Pincus was impressed by the quality of the work done on the island, and he was encouraged to learn that dozens of birth-control clinics were in operation. "I came to the conclusion that experiments could be done in Puerto Rico on a relatively large scale," he wrote. He thought it would be fairly easy to get between one hundred and three hundred women "with some intelligence" to cooperate in a series of experiments. The same experiments, he said, "would be very difficult in this country."

Pincus had made the trip to San Juan to lecture a group of scientists on "Biological Synthesis and Metabolism of Steroid Hormones," but he took the opportunity to rest and get some sun. He'd been working too hard, and feeling worn. It concerned him enough that he went to see a doctor, who ran tests that showed Pincus's white blood cell count was "somewhat elevated." When Pincus wasn't sunning on the beach, however, he spoke to several doctors about the state of birth control on the island. His host in Puerto Rico was Dr. David Tyler. During World War II, Tyler and Pincus had collaborated on a study into the adrenal gland's function in fatigue. Tyler was now the head of the Department of Pharmacology at the University of Puerto Rico. It occurred to Pincus that with a qualified scientist like Tyler supervising, it might be possible to do high-quality scientific research on a population desperate for better birth control.

McCormick was skeptical, for reasons perhaps suggesting racial bias as well as class bias. She doubted whether Puerto Rican women could be trusted to follow the rigorous testing regimen, and she wasn't sure the doctors and nurses in Puerto Rico were qualified to run the tests and record results. On the other hand, she believed in Pincus, and she wanted his work to move more quickly.

※

Pincus had not mentioned Puerto Rico casually. He had settled on the Caribbean island, an unincorporated territory of the United States, after considering India, Japan, Hawaii, Mexico, New York City, and Providence, Rhode Island. Tyler wasn't the only American working on the island; there were many others. That meant the language barrier would be low and professionalism high. Flights between the United States and Puerto Rico were frequent and often direct. Best of all, birth control had been legal in Puerto Rico since 1937.

Pincus was ready to push ahead.

Like McCormick, however, John Rock was concerned. He wanted "ovulating intelligent" women who could be relied upon to carry out his instructions, and he wanted to see firsthand how they responded to the experiments. But even if they acquired sufficient numbers of women in Boston, Rock wasn't sure testing would ever work. The only women who had consistently stuck with the onerous procedures to that point had been his infertile patients, who were motivated to follow doctor's orders in the hopes of getting pregnant. Rock believed that working women and women with children would never have the time to submit to all the urine tests, Pap smears, and endometrial biopsies required.

Pincus had a different view. Women in Boston might indeed be more reliable, and infertile women might be highly motivated, but he and Rock would never get enough of them to participate. Puerto Ricans, on the other hand, appreciated the gifts bestowed by Americans, or so many Americans of his generation liked to think. The island had long served as a test tube for social scientists, never more so than in the years after World War II. And overpopulation and poverty had long been serious issues on the island. Puerto Rico was poor and crowded, and family sizes were large. Between 1940 and 1950, the island's population had grown 18 percent, to 2.2 million, and its urban population had jumped by an astonishing 58 percent. The fertility rate in Puerto Rico was 17 percent higher than in seven other Latin American countries, and 34 percent higher than the United States. By the time she was fifty-five years old, the average mother in Puerto Rico had borne 6.8 children. There was a sense of urgency about the rapidly growing population, and even though Puerto Ricans were predominantly Catholic, they were already familiar with birth control. Leaders of the eugenics movement had long promoted campaigns for voluntary sterilization among Puerto Rican women, funded largely by Sanger's friend, Clarence Gamble. Gamble had been working on birth control almost as long as Margaret Sanger, although he had done so more quietly.

Gamble was born in 1894 in Cincinnati. His father, David Berry Gamble, was one of the last in the family to play a key role in the family company, Procter & Gamble. Clarence Gamble was a Harvard Medical School graduate who became a professor of pharmacology and decided to devote his great wealth to the cause of birth control. Though they often worked together, Gamble and Sanger did not always agree. While Sanger wanted to build the birth-control movement around doctors, believing it would add legitimacy to the cause, Gamble understood that most women did not have private physicians and would never visit a birth-control clinic. He wanted every child born to be a child desired by responsible and qualified parents, and he believed that the only way to make that happen was to find a cheap, simple contraceptive that could be distributed without doctors and nurses getting involved. Gamble wanted to re-engineer the world, making it safe for his kind of people—white, prosperous, and hard working—by reducing the fertility of those he deemed less desirable.

In Puerto Rico, thanks to the work of Gamble and others, women who delivered their babies in hospitals often requested sterilization before being discharged. One American doctor working on the island said she was "horrified" at how common the sterilizations were in the 1950s. "It was a habit," the doctor said. "I mean, it was a thing that they did . . . and the women were determined not to have more children." The more the Catholic Church complained about the sterilizations, the more Puerto Rican women tended to request the procedure. The Church was only helping to spread the news of the procedure's availability. To Pincus, that meant Puerto Ricans might reject or ignore the Church's instructions and embrace a birth-control pill—assuming, of course, that it worked. In the end, both Rock and McCormick agreed to try it Pincus's way.

They were about to launch one of the boldest and most controversial field trials in the history of modern drugs.

SEVENTEEN

A San Juan Weekend

As Pincus began planning Puerto Rican trials for the birth-control pill in 1954, another scientist, Jonas Salk, was launching the first trials of his polio vaccine. At first glance, the men had little in common.

Pincus was working somewhat clandestinely on a project that was almost certain to stir controversy and provoke fear, a project that he could not have dared undertake on American soil. Salk was a national hero out to vanquish a common enemy. The nation was riveted by his quest. Polio was killing and crippling American children at a rate of 50 per 100,000. It was a disease that hit without warning and seemingly at random, leaving its survivors in wheelchairs and leg braces. Even when polio cases peaked in the 1940s and 1950s, ten times as many children were killed in car accidents. But conquering the virus became a national priority of the highest order. More than two-thirds of all Americans donated money to the March of Dimes, a charity formed to fight the disease. In 1954, more Americans knew about Salk's clinical trials than knew the full name of the president of the United States, Dwight David Eisenhower. There had never been a medical experiment like it.

Salk would spend tens of millions of dollars on his vaccine trial, while Pincus would operate with a first-year budget of less than twenty thousand dollars. Salk would go on to test his drug on six hundred thousand children during field trials, while Pincus was hopeful that he would round up three hundred subjects.

But Salk and Pincus were not as different as they might have seemed. Both men were willing to overlook potential risks for the sake of speed. Both men were willing to step outside the cloistered world of academic research and engage with advocacy groups that sought to guide and hurry along their work. Finally, both men were unconcerned with profiting from their inventions. In 1954, as Pincus negotiated with Planned Parenthood for support and funding and arranged with Searle and Syntex for supplies of the necessary progesterone compounds, he too focused on the scientific work ahead, not on the money. As he said to the leaders of Planned Parenthood and to members of his own Foundation's board, he was operating on the assumption that the drug companies producing the compounds would hold patent rights if his experiments did, in fact, produce a birth-control pill.

※

Not since the 1930s, when he had been dabbling with *in vitro* fertilization, had Pincus been so close to a discovery that had the potential to change human lives. And while the public clamor for birth control hardly equaled the clamor for a polio vaccine, the growing concern about population growth lent legitimacy to the work of those studying human reproduction.

"The earth's population will double to 5,000,000,000 in the next 70 years if the present birth-death rate continues," warned a United Press story in what would turn out to be a conservative estimate. While birth-control campaigns and legalized abortions were reducing the rate of population increase in some of the world's more affluent nations, Asian and Latin American countries were still growing more rapidly than ever. At the same time, penicillin and other medical innovations were helping people live longer.

In 1954, the first United Nations Population Conference was held in Rome, with representatives from almost every country attending, prompting still more headlines in the American press. In the United States, where land was abundant and the economy strong, few people worried about overcrowding. But there was a great deal of concern in Puerto Rico, and that concern was becoming a topic of conversation in the United States, especially in New York City, where Puerto Rican immigrants were arriving in extraordinary numbers. Puerto Rico's population doubled between 1900 and 1950, making it one of the most densely populated countries in the world. The island was twelve times more densely packed than the United States, and the crowding would have been even greater if so many Puerto Ricans were not fleeing to American cities such as Chicago, Philadelphia, and especially New York. In 1950, New York City was home to 246,000 Puerto Ricans. Three years later, the number was 376,000 and still rising. By 1953, nearly one in every ten residents of Manhattan was Puerto Rican. The arrival of so many Puerto Ricans in the United States helped bring attention to problems on the island, where 2,245,000 people lived in an area one hundred miles long and thirty-four miles wide. As Puerto Ricans left home for better jobs and better lives, the crowding became more than a Puerto Rican problem; it became an American problem, and perhaps a sign of things to come.

In Puerto Rico and many other developing countries, birth rates were holding steady but death rates were declining. Malaria was being wiped out. Water supplies were improving, which meant fewer deaths from intestinal diseases. And new drugs were stopping the spread of tuberculosis. But better health did not translate into a better economy. Despite a massive American push that opened hundreds of new factories on the island after the war, one in six Puerto Ricans remained unemployed, and the industrialization pushed more people to the island's cities, only heightening the sense of crowding. The issue was so pressing that 8.5 percent of married Puerto Rican women under the age of fifty had volunteered for sterilization. There were no reports on how many men underwent vasectomies, although the number was certainly much smaller. Courts sometimes ordered the sterilization of men who were deemed insane or criminally dangerous, but few men volunteered for the procedure in order to prevent pregnancy. Responsibility to control reproduction fell to the women, and women were taking action as best they could. In a 1952 study, 80 percent of women said the ideal number of children for a woman was less than four.

"There is no advantage in having more children," one Puerto Rican woman said.

> *If one is poor one shouldn't have more than two. The rich can have more because they have the money to educate them and are not sacrificed or even killed working as the poor do. . . . The rich care better for the sons, but it is great work for the poor to rear them, and the wife of the poor gets sick with many children because she can't feed herself well nor have the medicines if she needs them. So, two is enough.*

A social revolution was underway in Puerto Rico. Poor women were doing everything they could to have fewer children, believing that there would be no escape from poverty otherwise. Investment on the island after the war had made more jobs available (mostly to men), but factory jobs still didn't pay enough to support families of ten. In 1951 and 1952, as Pincus was beginning his progesterone research in Worcester, a young sociology student at Columbia University, J. Mayone Stycos, went to Puerto Rico to interview women for his doctoral dissertation. At one point he asked some of the women if they were quick to have children after marriage because they were afraid of being branded *machorras*, or barren women, in Stycos's translation. Their response surprised him. Most of the women said they didn't care, and five out of fifty-six said they would be happy to be infertile.

"In older times," one woman said, "women were ashamed not to have children. . . . Women of today even expose themselves to death trying to avoid children."

Some of the men in Puerto Rico told Stycos that they were sleeping with other women to avoid getting their wives pregnant after their families had reached what they considered desirable sizes. "One has to look for other women," one man said, laughing. "That is the most suitable method, and the handiest." He went on to say his wife was jealous but that when he explained his logic, "she made up her mind and agreed with me."

But not all women were so compliant. One told Stycos about her own method of birth control. "You know what I do?" she asked. "I hit him with my legs and throw him out of the bed. We have quarreled about it for nine years and he even threatened to kill me."

Some women said the fear of pregnancy had crushed their desire for sex. "How could I enjoy it, thinking of what would become of me if I have another child?" one woman asked. In a few cases, women said they intentionally married men who were rumored to be sterile because they preferred not to have children. One or two children might have been acceptable, perhaps even three or four, but the women felt as if they had no choice. Once they married and began having sex, they feared they would wind up with more children than they could handle, despite the fact that Puerto Rican women had greater access to birth control than most American women.

A court decision in 1937 had legalized the distribution of birth control for medical purposes, and in response about 160 clinics had opened across the island by the 1950s. The devices handed out didn't work well, often because the women used them only intermittently.

Between 1945 and 1950, condom distribution increased by 50 percent, and Stycos found that 72 percent of families practiced birth control at some point. Four percent of the women he interviewed admitted having one or more abortions, but Stycos suspected that the abortion rate was actually much higher because women were reluctant to discuss the subject. Though abortion was illegal, it became so commonplace in the 1950s that Puerto Rico developed an international reputation as a place where the procedure could be obtained, no questions asked. For Americans, it was a quick enough flight yet far enough away to offer anonymity. They called it a "San Juan Weekend," with an appointment at the abortion clinic on Friday and a return flight home on Monday. The whole thing, including airfare and hotel, would cost about six hundred dollars.

Despite the high rates of abortion and the widespread acceptance of birth-control devices, Puerto Rican birth rates declined only slightly in the first half of the twentieth century. Women in rural areas had an average of 6.8 children and women in urban areas had an average of 4.8. Stycos believed that concern over pregnancy wasn't translating to action for many women, or at least not consistent action. Some men and women said they were reluctant to use birth control because the Catholic Church rejected it, but most offered more practical excuses. Men and women both complained that condoms were not always handy when needed. Many men refused to use condoms because they didn't like the way they felt, and some women rejected them because they had heard rumors that they might cause cancer or hemorrhages. Men said they were unconcerned about birth control because it was the wife's responsibility to raise the children no matter how many were born. One man told Stycos that when he felt his wife had too many children, "I will have her sterilized."

Why did birth control efforts in Puerto Rico fail? Stycos cited male dominance in Puerto Rican society, poor communication between spouses, modesty among women, and misinformation about how birth control worked. He complained of a shortage of personnel at health clinics, apathy among doctors, and the failure of government to educate women about their options even in a country with one of the most ambitious and widespread birth-control programs in the world. Rapid population growth in Puerto Rico, he said, "imperils the whole society," and nothing short of widespread social change would save it.

But what about a miracle drug?

On the final page of his report, Stycos noted that scientists were reportedly working on an oral contraceptive and that some believed it could be a panacea for the world's population problems. The young scholar was skeptical. The problems in Puerto Rico ran too deep, he said. No miracle drug could change the way people lived and loved, especially in a place as poor as Puerto Rico, he argued.

※

When Pincus visited the island in February 1954, he met Dr. Edris Rice-Wray, medical director of Puerto Rico's Family Planning Association. Rice-Wray was not a Puerto Rican; she was a native of Detroit who earned a bachelor's degree from Vassar College and a degree in medicine from Northwestern University. Those meeting her in Puerto Rico might have mistaken her for the wife of a doctor or diplomat. She wore her reddish-brown hair in a carefully curled bouffant and adorned her practical suits and dresses with pearl necklaces and glittering brooches. But Dr. Rice-Wray's appearance was the only thing conventional about her.

"I remember when I was in medical school saying to the boys, 'Well . . . I'm going to get married, I'm going to have two children, but I'm going to practice medicine.' And they said, 'Well, maybe you'll get married and have two children, and maybe you'll practice medicine, but you won't do both.' "

She did both.

Rice-Wray was an active member of the Bahá'í Faith, a religion founded in the mid-1800s that teaches there is only one God and one human race, and that all of humanity will one day come together in a peaceful society. Rice-Wray pursued her education and career in Puerto Rico in large part because she believed she could help fulfill the mission of her religion. But there were also practical reasons behind her decision to get involved in contraceptive work. During her second year of medical school, she got married and visited a Planned Parenthood office to get advice on how best to avoid pregnancy. She told the volunteers there that she did not want children to come too soon and ruin her chances of becoming a doctor. They recommended a diaphragm.

After graduation, she had two girls and began working as a physician in private practice on Chicago's North Side. Once or twice a week she would work at health clinics for the poor, where the problems women faced came as a revelation. "You know, women with five or six children, poor, with a husband who's very demanding, and has a terrible life. You're stuck. You're in a trap," Rice-Wray reflected later. Even with her medical practice and relative prosperity, Rice-Wray felt trapped eventually, too. "If you know about suburban life," she said, "you take your husband to the train, and you put him on the train, and you . . . take the kids to school and you . . . go get the laundry . . . and that's what the wives were doing . . . and, well, you know, it wasn't enough for me." Rice-Wray divorced her husband and decided she wanted to see more of the world than suburban Chicago, and that she wanted to do something to help the other women who felt trapped. She'd been taking Spanish lessons, and a friend from Puerto Rico offered to write a letter on her behalf to an official in the Puerto Rican health department, the Departmento de Salud, recommending Rice-Wray for a job. In 1949, she moved with her children to San Juan and went to work.

A few months prior to Pincus's visit to Puerto Rico, Rice-Wray and another official at the Puerto Rican Association of Population Studies, Rafael Mendez Ramos, had written an impassioned letter to William Vogt at the Planned Parenthood Federation of America, pleading for help. "People probably wonder what Puerto Ricans are doing about this situation," they wrote. "The answer is, first, that the people are gradually becoming conscious of their problem; second, that they are doing nothing about it."

There was talk of creating more jobs, and, of course, there were planes leaving daily for New York City, but it wasn't enough, they complained in their letter. Dozens of public health clinics were operating across the island, with ample funding from the government, yet neither the government nor the doctors and nurses operating the clinics were interested in promoting birth control. Political and religious pressure kept the government from doing more, but there was nothing stopping private groups like Planned Parenthood from stepping in, Rice-Wray and Ramos said. They argued there were good reasons for Planned Parenthood to get more deeply involved in Puerto Rico. For one, researchers on the island were already engaged in a comprehensive study of birth-control habits, interviewing more than a thousand women on why they chose not to take advantage of birth-control clinics, why they frequently stopped using birth-control devices, and what could be done to help them make better uses of available resources. Once the study was complete, the researchers planned to launch a campaign to promote greater use of the clinics and the birth-control methods they provided.

But Rice-Wray worried that sociologists conducting the studies were not equipped to carry out a public education campaign. For that, she wanted Planned Parenthood's help. She proposed making her organization, the Puerto Rican Association of Population Studies, an affiliate of Planned Parenthood. "We have 160 clinics on 3500 sq. miles of land and yet the results have been disappointing," she and her colleague wrote. But the clinics were all in place and education campaigns had already been launched. With only a little more money and effort, Puerto Rico could become ground zero in the war on unwanted pregnancy. "Such a program would surely have world wide significance," they wrote, "for it could point the way for programs in other highly populated under developed areas."

When Rice-Wray met Pincus, she had the feeling he was desperate, "looking for anybody." Rice-Wray was cautious. "I was kind of scared at first really, because I had never heard of a pill," she said. "But he convinced me. He's a very persuasive fellow." Once convinced, she urged him to move fast. "Our great opportunity," she said, "is now."

※

For Pincus, the choice came down to practicalities. Margaret Sanger wanted him to enlist women from her New York birth-control clinic, known as the Margaret Sanger Research Bureau, where Dr. Abraham Stone would supervise the testing. But Pincus wasn't sure women in New York would put up with the strenuous tests required. He never explained in his letters why he thought Puerto Rican women would be more compliant than New Yorkers. Maybe it was easier to imagine them going along with the program because they were poor, or because they were farther away, or because they were desperate for a better way to control the growth of their families. Another reason: The island was crowded, which meant he had more potential subjects from which to draw.

Pincus didn't discuss it much, but he knew that the women enrolled in future progesterone trials would have more than annoying medical procedures to complain about. At the same time that he was scouting locations for the next round of experiments, he and Rock were tallying the results of their tests on infertile women from Boston and Worcester. Of the seventy women who participated over the course of the first year of those small-scale, somewhat informal trials, fifty-six women experienced a change in the pigment of their nipples; fifty-three suffered breast soreness; thirty-seven said their nipples became enlarged; forty-four experienced nausea, vomiting, or both; fifty experienced vaginal discharges; twenty-three endured increased urination; twenty-one reported increased libidos; four reported reduced libidos; and four experienced lactation. Only five reported no side effects.

Even so, side effects were low on Pincus's list of concerns. No doubt the fact that he was a man and had never been pregnant contributed to his callousness. On the other hand, Sanger and McCormick were not raising concerns about side effects, either, and Pincus was planning to include several women, including Dr. Rice-Wray, on his team. It's more likely that Pincus was, as usual, trying to take a practical approach. There would be no point worrying about side effects if he couldn't get women to participate in the trial. Also, he was still experimenting with different compounds. That meant he might yet find a form of progesterone that worked at lower doses and produced fewer unwanted reactions. The thing to do was to keep trying.

On October 19, 1954, Pincus wrote to Dr. Manuel Fernández Fuster, one of the doctors he'd met at the University of Puerto Rico, asking him to "assemble a group of 50 women." In a memo to Rock, Pincus cautioned that suitable subjects for the experiment "must be informed, intelligent, cooperative females who give evidence of normal ovulation." Women with children, he observed, might be too distracted to take their temperature readings at precisely the same time each day. He added that the ideal subjects would be women for whom "pregnancy would be acceptable, or at worst inconvenient," in case the pills failed.

To keep the work simple and boost chances of cooperation, they would begin with nurses and female medical students, and they would ask them only for daily temperature readings. Pincus was learning. He knew it would be difficult to find women willing to take big doses of progesterone and submit to daily urine tests as well as temperature readings. By seeking only temperature readings, he would find out whether the university was capable of recruiting women and whether the women could be counted on to show up every day for their tests. And if the women did cooperate, the temperature readings would tell him if they were ovulating regularly. The women who stuck with the program and ovulated regularly would then be recruited to join the clinical trial.

But Pincus wasn't counting on the Puerto Rican women alone. He had a backup plan—another group of women available for testing—and he found them close to home, in Worcester. These women would not be motivated in the same way the patients of Rock and Kirkendall were by a desire to get pregnant; nor would they be inspired as the nurses and medical school students in Puerto Rico were by an interest in furthering medical research. The women Pincus had in mind were even better subjects in at least one respect because they would have no choice but to participate.

EIGHTEEN

The Women of the Asylum

Once, Florence Kouveliotis had been a great beauty—petite, brown-haired, brown-eyed, with a smile that flashed and warmed. But by 1953 she was a wreck.

Genetics had done some of the damage, but men and marriage and childbirth had taken their toll, too. Kouveliotis's parents had both showed signs of severe mental illness, although neither of them was formally diagnosed. Their instability meant that Florence, the oldest girl, had to take care of her five siblings from the time she was still a schoolgirl. Once she reached her teens, the cracks in her own sanity were beginning to show. She became paranoid and delusional. She grew deathly afraid of germs. She found work and managed to hold a job stitching shoes at a factory in Lowell, Massachusetts, but maintaining a normal routine became increasingly difficult. By the age of twenty-nine she was still unmarried, which was unusual for a woman at the time. Perhaps she knew she was unstable, or perhaps men recognized it once they got past gazing upon her pretty features. But when she was thirty, some of her relatives arranged a marriage with a bridge painter from Greece who needed an American wife to avoid deportation. Florence quickly became pregnant. Her husband drank heavily and traveled in search of bridges to paint, leaving Florence at home to raise their son. Rather than calming her turbulent mind, motherhood made her more anxious than ever. To a woman with a phobia of germs, a baby was horrifying, a crawling, drooling incubator of filth and disease. She tried everything, even bathing him with bleach, but nothing helped. Unable to cope, she often left him alone, unwashed and unfed.

Before anyone noticed that she was struggling, Florence was pregnant again. In 1949, when she went to the hospital to deliver, she brought her two-year-old boy with her. Hospital officials, seeing that the child was covered in sores, immediately removed the baby from her custody and helped arrange for adoption by a relative. But her second and third children were not so lucky. Florence heard voices telling her to kill the children, and several times she tried. When they got older, the children ran away. They told their teachers on several occasions how they were being treated at home and at least once told police. Finally, in 1953, her two youngest children were placed in orphanages and Florence was sent to the Worcester State Hospital for the mentally ill. She was diagnosed with paranoia, schizophrenia, and multiple-personality disorder, according to her daughter, Tina Mercier. Doctors treated her with insulin therapy and at one point performed a lobotomy, although her family would not learn about it until decades later, when an X-ray revealed the telltale indentation in her skull. She was also used as a subject by Gregory Pincus and his staff for some of the earliest tests of progesterone as an oral contraceptive, a fact that Tina, some sixty years later, would find sadly ironic. A birth-control treatment, she said, is the one thing that might have done her mother some good—if it had come along earlier.

"She never should have had any of us," Mercier said.

※

The asylum was more than a century old. It looked like a medieval fortress, constructed of four-foot-square stones that glowed bloodred in the sunset. A giant bell tower rose three stories higher than the rest of the imposing structure. Iron bars covered the windows. Visitors passed through a great iron gate and drove along a twisted path among gnarled trees, past cottages for the senior staff, and up a steep hill before finally reaching the baroque administration building. A grand stairway led to the administration's main entrance. Two clinical wings extended from either side of the main building—one for men, one for women. Each wing had a secondary wing set back from its neighbor, and behind each secondary wing was yet another. The design was maddening, resembling the endless labyrinths of the mind.

In 1927, Katharine McCormick had established the Neuroendocrine Research Foundation at Harvard to fund research she hoped would lead to a cure for her husband's severe mental illness. Harvard at the time had used McCormick's money to conduct studies among mentally ill patients at the asylum in Worcester, and Hudson Hoagland had been one of the Harvard scientists assigned to the project. Back then, there were no laws banning the use of mental patients in scientific experiments. So when Hoagland and Pincus set up their foundation in Shrewsbury years later, they viewed the local asylum as a convenient resource. In fact, they conducted experiments there so often that the Foundation maintained a laboratory in the asylum's main building and kept scientists there year-round.

By the 1950s, the hospital housed three thousand patients and one thousand staff members (though only about thirty doctors). The man in charge was an enormously overweight Harvard grad named Dr. Bardwell Flower. "Dr. Fartwell Blower," the staff called him behind his back. The food was so poor that patients in the rear wards who received little sunlight were often treated for scurvy. The stink of excrement and disinfectant slapped visitors in the face. A single nurse or attendant cared for one hundred patients at a time. Security was lax. Worcester officials congratulated themselves for seldom using restraints on patients, but sometimes their attempts at leniency backfired. In 1943, one inmate beheaded another. And it was not unusual for patients to bind and gag their doctors or nurses and walk out of the hospital. By the mid-1950s, doctors were beginning to experiment with chlorpromazine for schizophrenia, but most still relied on shock treatments, lobotomies, and barbiturate sedatives. The treatment of mental illness remained a rough and imprecise science, one that outsiders found deeply disturbing to witness. Dr. Enoch Callaway recalled that one patient with a lobotomy worked as a maid. Once, when he and his wife briefly resided in one of the hospital's apartments—"the honeymoon suite," they called it—the lobotomy patient came in and tried to change the sheets while the bed was still occupied. But most moments were not so amusing. Dr. Callaway watched as "warehouse-like rooms" filled to capacity with "patients who have defeated our best efforts." In some ways, he said, it was a model mental hospital. In other ways, it was a snake pit.

It was not difficult for Pincus to obtain permission to experiment on patients at the asylum. No permission slips needed to be signed. It would appear that Dr. Flower was simply pleased to have a few more physicians on hand in his enormous hospital. It helped, too, that McCormick offered money to paint and refurbish some of the asylum's wards in exchange for cooperation in the progesterone study. And so they began administering progesterone and estrogen in varying doses to women diagnosed with paranoia, schizophrenia, melancholia, manic depression, chronic alcoholism, Alzheimer's disease, Pick's disease, and more.

Not all the women at the Worcester State Hospital were chronically ill, however. All over the country in the 1950s, mental hospitals experienced a population explosion of their own. Ironically, some of the women who voluntarily checked into the hospital were those who might have been most helped by an oral contraceptive such as the one Pincus was testing. They were women suffering from anxiety or depression, conditions prompted or exacerbated by too many pregnancies or the strains of taking care of large families. Some women were not necessarily mentally ill but crumbling under the pressure of abusive husbands or domestic conditions they could no longer bear. Mildly ill patients received counseling, rest, and barbiturate sedatives. The more seriously ill, including women such as Florence Kouveliotis, were treated with electroshock, insulin shock, and lobotomies. Chronic cases who didn't respond to treatment at the Worcester State Hospital were steadily moved to higher floors and more distant wings of the hospital until they made it to the second and third floors of the third wing, which the staff referred to as "the back wards," giving profound new meaning to the expression "Out of sight, out of mind." Women on the back wards were issued so-called "strong dresses," garments made of heavy blue-gray canvas with reinforced stitching on the hems, sleeves, and neck. No belts or ties were provided. They all had their hair shorn to shoulder length. "Since we had no effective and safe sedatives," Dr. Callaway wrote, "most of them milled about in aimless agitation, defecating and urinating as the urges arose." Many of the women were disheveled and malnourished. Some of them would fling themselves at guards or fellow patients, biting, scratching, and pulling hair in frustration. When he was assigned to perform an annual review on one of the women in the back wards, Dr. Callaway said, "I could not help imagining her as a young woman, full of hope for the future, and never dreaming that she would end her days waiting for death in a warehouse full of broken minds."

The back wards offered one of the most miserable collections of human beings imaginable, but they also supplied a nearly endless array of patients for research projects like the one Pincus was planning.

"You could do experiments there you would never think of doing these days," Callaway recalled. "You never asked the patients for permission. Nobody supervised. We gave amphetamines to schizophrenics once. I'd say, 'John, would you mind if I gave you a shot to see what it does to you?' And he'd say, 'Oh, sure doc.' The amphetamines just made them more talkative."

Though the patients were accessible and largely cooperative, they were still not ideal subjects. Conditions in the asylum were so dire that Pincus's team of researchers feared for their safety. His section heads sent a memo titled "SUBJECT: GENERAL LABORATORY CONDITIONS," which read, in part:

> *[W]e wish to inform the directors that the working conditions in these laboratories after 10 P.M. when the night watchman finishes his rounds are not safe. The laboratories are easily accessible to patients even when all the doors are locked. Since it has been the practice of some people at the technical level to work at night, it is felt by us, being responsible for their welfare, that some directive should be forthcoming to establish the conditions for such practices.*

Pincus managed to enroll sixteen women—all of them classified as psychotics—for his first round of progesterone tests. He also gave the hormone to sixteen men to see how it would affect their sterility—and, while he was at it, to see if it might help with their mental illness.

Unfortunately, these tests did not prove helpful. The women were not having sexual intercourse, which made it difficult to be certain they weren't ovulating. It was similarly difficult to tell whether the compounds affected ovulation because psychiatric problems disrupted the menstrual cycles of many of the women. As for the men, Pincus believed that the progesterone might lower their sperm counts and reduce their sexual desire. It's not clear if he was interested in this for purposes of birth control or merely for curiosity. Sanger and McCormick had made it clear that they weren't interested in a birth-control pill for men. They didn't trust men to take the responsibility, and they wanted women to possess control of their own bodies and their own fertility. But even if progesterone failed as a contraceptive for men, or if his sponsors were uninterested in such a thing, Pincus thought there might be another useful application. If progesterone subdued a man's libido, the hormone might be administered in combination with psychotherapy to cure men of homosexuality. In the end, though, his tests were inconclusive. One psychotic male subject appeared to have shrunken testicles after taking progesterone for five months. Nurses said the same patient became more "feminine" while taking the hormone. The investigation went no further in part because the researchers couldn't persuade the male asylum patients to produce semen samples. They did observe the frequency with which the men masturbated but found no change. They also concluded, according to Pincus, that the men taking the progesterone "were just as psychotic as they were when we began to give them the drug."

With no more time to waste, it was back to Puerto Rico.

NINETEEN

John Rock's Hard Place

FOR JOHN ROCK, the time had come to make a decision.

Was he in? Was he going to commit to a project that would put him in direct opposition to the Catholic Church, the same church to which he had been devoted since boyhood?

Rock was not only a practicing Catholic but also a respected medical doctor with a thriving practice and a spotless reputation. What's more, he was sixty-four years old and nearing mandatory retirement from Harvard, and he had begun thinking about launching an independent practice. Pincus, Sanger, and McCormick were all gambling by getting involved in the search for an oral contraceptive, but they were outsiders with little to lose except their time or money. Rock had more on the line.

By 1954, he had seen enough of Pincus to know how the scientist operated and to know that it made him uncomfortable. Rock believed there would be no harm in giving women heavy doses of progesterone for a few months, especially when it offered infertile women hope of becoming pregnant on the rebound. But he had doubts about putting women on the hormone for prolonged stretches. The potential side effects were serious. The regimen was taxing. And no one knew what kind of long-term health or reproductive problems might arise. He reminded Pincus, as well as McCormick and Sanger, that the work should not be rushed.

McCormick, ever anxious to push forward, grew frustrated at times with Rock's conservative approach to science. "Dr. Pincus is imaginative and inspirational," she wrote. "Dr. Rock is informative and very realistic about medical work." McCormick was astonished at the difficulty of the testing and marveled that any women would agree to it. She suggested paying participants to cooperate, but Pincus and Rock refused. What they needed for the tests, McCormick said, perhaps facetiously, was "a cage of ovulating females." Rock explained to McCormick that his patients—and other patients, as well—were not to be exploited. These were "people who want babies," he said. They were emotionally fragile and needed tender care. They were also young and looking forward to long lives as mothers and grandmothers. He would not do anything that might damage their health "for just an abstract research thing."

Sanger didn't trust him. Yes, she admitted, he was charming and handsome. And yes, Rock had shown commitment to the cause when he had agreed to enroll his patients in clinical trials with Pincus. Still, Sanger refused to believe that a Catholic gynecologist could ever truly dedicate himself to the cause of contraception. "Abram, Abram," she wrote to Abraham Stone when she learned that Planned Parenthood had appointed Rock chairman of one of its scientific research committees, "What has happened to you to allow the enemy to walk in the front door?" If Pincus had not come to Rock's defense, Sanger might have shoved Rock off the team. But Pincus needed Rock to give him legitimacy in the medical community.

Another concern was Rock's health. He had suffered a major heart attack in 1944, followed by several smaller ones. Two years after the first attack, a car accident had killed his eldest child, Jack. Over the course of the 1940s, his wife had suffered a number of illnesses that had left her partially paralyzed. Through it all, Rock pressed on with his work. He eliminated salt from his diet and gave up smoking tobacco, stuffing his pipe with corn silk instead. Perhaps the health scares and personal tragedies freed him to take more chances and do what he believed was right. Here was a doctor who had devoted most of his career to fertility treatment who was now beginning to focus his time and energy on the opposite side of the issue—contraception. Did that make him a rebel? Rock did not see it that way. He was interested in solving the central problems surrounding human reproduction. In Rock's mind, there was no reason to be afraid of contraceptive research; it was a close relative to fertility research. Nor was there any reason to be afraid of discussing sex.

Earlier in his career, Rock had been too busy delivering babies to explore the mysteries of human reproduction. Only after his first heart attack did he begin to reduce his patient load and spend more time on research. Initially there had been little money for research. Patients paid the bills. Now, as one of the top figures in his field working at one of the top universities in the world, grant money flowed from drug companies, foundations, and the federal government. Fertility clinics, once rare, were cropping up around the country as women in the 1950s tried to join the postwar rush to get married and pregnant. But even the doctors running the clinics did not fully understand their field because it was so new, and they often turned to Rock for advice. Most of his time was still dedicated to helping infertile couples try to have children, but the work was difficult. Trained as a surgeon, Rock tried delicate operations to repair fallopian tubes, but the results were rarely successful. When a man's low sperm count was responsible for the infertility, Rock tried freezing the sperm and weeding out the sluggish ones. He tried spinning semen in a centrifuge to boost its concentration. He told men to eat more lettuce for vitamin E and to lose weight because lean animals tended to be more fertile than heavy ones. He prescribed large doses of testosterone, thinking that it might work the way progesterone did on women—shutting down their reproductive systems and giving them a temporary rest. He tried everything he could think of, but most of the work had little effect. By the 1950s, he admitted to colleagues that he and his fellow researchers knew "pitifully little that was of practical value."

※

On May 6, 1954, Rock was the featured speaker at Planned Parenthood's annual luncheon in New York City. Before the event, Planned Parenthood officials expressed concern: What would the press say if they found out that their keynote speaker was Catholic? Rock had this response: "Just tell anybody who asks the point in question that I am a Roman Catholic, and then you might add 'and what of it.' Being one does not blind one to the paramount importance of the world population increase."

In his speech, Rock managed to be both charming and challenging. He began:

> *I take it you are here today because you are fascinated by the subject of babies, or because you ought to be; and I am given the privilege of addressing you because I know something about the business of having babies—or because I ought to. This most ancient of human endeavors has allured me as a gynecologist for thirty years, and I must confess, under other guise, for nigh on fifty.*

He went on to say that the audience was probably wondering whether he was for or against having babies. Rock said he was for it, strongly—and he was not only in favor of babies but also in favor of sex. The desire for sex—"this impulse that forces Adam to present, and Eve to receive"—was something humans should cherish, not repress. In other mammals, the sex drive lives only in the "lower nervous centers," he explained, but in humans it passes upward from the groin to the base of the forebrain, where the urge for sex becomes "inextricably blended" with something we call love. Without this instinct for love, he said, man would be no more than beast. He asked: "Why, if sex is so natural . . . should any married person want or be required to restrict either its exercise or its output? The gruesome answer is inescapable. . . . If fertility is uncontrolled, normal expression of the sex instinct, even with the monogamous family that evolved for this very purpose, results in the fusion of more spermatozoa and eggs than most parents can safely culture, or Earth care for."

The solution, he said, was not for men and women to stop having sex. That impulse was too strong to be denied. And even if married couples did try to suppress their urges, without sex love would grow cold. But there was a way to help married couples continue to enjoy sex without having more babies than they desired. The solution, he said, was to find "a contraceptive as effective as continence but without its inherent disadvantages."

Such a contraceptive would not be discovered by accident. It would require investment, he said. In 1953, the federal government spent thirty million dollars on efforts to control hoof-and-mouth disease, Rock told his audience. It spent another two billion dollars on nuclear weapons research. He concluded: "If we could muster just one-thousandth of this amount to finance the study of human reproduction, we would assuredly obtain the greatest aid ever discovered to the happiness and security of individual families—indeed of mankind. This would avert Man's self-destruction by starvation and war. If it can be discovered soon, the H-bomb need never fall.

"The urgency is on us: ample talent is available. Let us mobilize it with dollars and devotion."

Rock might have supposed that he was preaching to the choir. If anyone could have been expected to join his call to action, it was officials and donors of Planned Parenthood. Certainly, the organization had the ability to produce two million dollars in funding.

But Planned Parenthood was no longer the radical organization it had been in its earliest years. Liberalism was out of fashion in the 1950s. The economic depression was over and American wages were rising, but poor and working-class families were making little progress. Labor unions were losing power, and leftists were coming under attack. The radical feminism of Margaret Sanger not only went out of style, it also became dangerous. When Senator Joseph McCarthy launched an investigation into the communists infiltrating American society, hysteria gripped the nation. Leaders of the birth-control movement, like so many other liberal activists in the 1950s, toned down their voices. Women of the 1950s were still expected to serve their men, and if a marriage failed, it was almost always the wife's fault. If a husband drank too much or carried on affairs, he was probably seeking refuge from an unpleasant home. If a woman failed to make her man happy, she wasn't trying hard enough. "Two big steps that women must take are to help their husbands decide where they are going and use their pretty heads to help them get there," Mrs. Dale Carnegie, wife of the best-selling self-help writer, wrote in *Better Homes and Gardens* in 1955. "Let's face it, girls. That wonderful guy in your house—and in mine—is building your house, your happiness and the opportunities that will come to your children."

But there were signs in the 1950s that women were ready to rebel against sexual and marital norms. In 1954, readers responded with outrage when movie star Marlene Dietrich wrote in *Ladies' Home Journal* that women needed to subordinate themselves to men if they wanted to be loved. "To be a complete woman," Dietrich said, "you need a master." She went on to say that women should wash the dishes and emerge "utterly desirable." One woman shot back: "Out here where I live, reasonably intelligent [married couples] . . . learn to live and work together."

According to the Kinsey report, 85 percent of all white men had had premarital sex, which meant, of course, that a roughly equal number of women were doing the same. Women were at least beginning to talk among themselves about opportunities beyond housewifery. Some of them were also discussing their frustrations with psychoanalysts, who adapted the work of Freud to suggest that sexual repression could damage a person's mental health. Divorce rates were creeping up. And women were being urged as never before to become active in their communities. In the 1930s, when *Ladies' Home Journal* urged women to help end the Depression, it suggested they do so by shopping more. By the 1950s, the *Journal* was suggesting they consider running for local office and working behind the scenes on political campaigns.

Even so, there was no young version of Margaret Sanger on the American scene to lead the revolution. No one yet dared declare that motherhood ought to be voluntary, that women had as much right to sexual pleasure as men, that marriage was not necessarily meant to be dominated by the husband, or that women were as entitled to a college education and good jobs as men were.

In 1949, *Ladies' Home Journal* ran a feature on the poet Edna St. Vincent Millay, along with a photo of her recently remodeled kitchen. "Now I expect to hear no more about the housework's being done," the magazine said. "For if one of the greatest poets of our day, and any day, can find beauty in simple household tasks, this is the end of the old controversy." But women weren't buying it, at least not all of them. In the mid-1950s, a housewife turned freelance writer named Betty Friedan began researching a book on a generation of women that had given up their dreams of emancipation for the security of big suburban homes with shiny modern appliances. Friedan claimed that male editors at magazines were out to convince women that housework provided all the self-expression and independence they needed. With her book, which she would call *The Feminine Mystique*, Friedan intended to show women another way.

In the first part of the decade, television shows made fathers the bosses. Only the men controlled the family's money and made the important decisions. But in 1955, *The Honeymooners* appeared on the air. Alice Kramden put her hands on her hips and told her husband Ralph that he was supposed to pay his lodge dues out of his allowance and she wasn't going to give him another dime until he learned to take better care of his money. In another episode, Alice informed her husband she was getting a job and *he* would have to start doing the housework.

Rebellion was brewing, and sex and gender were at the center of it. "Sex is something I really don't understand too hot," said Holden Caulfield, the protagonist of J. D. Salinger's novel, *The Catcher in the Rye*, published in 1951. "You never know where the hell you are. I keep making up these sex rules for myself, and then I break them right away." On July 5, 1954, less than two months after Rock's address to Planned Parenthood, Elvis Presley recorded his first single, "That's All Right." The song had a throbbing beat. As he sang, Presley straddled his microphone stand and gyrated his hips in a manner that thrilled young women and shocked their parents. His band, Elvis said, was "wearing out britches from the inside."

※

In 1954, Hugh Hefner was married, living in a lovely apartment in Hyde Park, and preparing for the arrival of his second child. He was also having an affair with a nurse who would soon help him make a sex movie. In the September 1954 issue of *Playboy*, Hefner had reprinted a medieval tale from Boccaccio's *Decameron* that describes the sex life of a gardener who is constantly being seduced by nuns. The same issue contained Jackie Rainbow posing nude for a centerfold, a short story about an automated sex machine that took the place of women in bed, and half a dozen photos of actress Gina Lollobrigida and her "generous bosom," but it was the Boccaccio story that earned Hefner a condemnation from the Church and a phone call from the chancellery. Despite the complaints, or perhaps in part because of them, *Playboy* was the fastest growing magazine in America.

※

Pincus and Rock were too old to be on the front lines of the revolution in sex that was shaping up, but that didn't mean they avoided it entirely. Goody Pincus never cheated on his wife, as far as his friends and relatives could tell. Rock was a great romantic, and as he grew older he became increasingly comfortable discussing sex in public. He liked to dance and drink and interview strangers about their sexual habits. One person who spent time with him in Puerto Rico remembered a handsome young man with a guitar serenading Rock from beneath the doctor's hotel window.

Pincus and Rock were old enough to recognize that a birth-control pill would not likely have much effect on their own lives, but they knew it had the power to unlock desire in others.

Even so, Pincus was focused on the science more than the possible consequences. He was more than three years into the project and still lacked a reliable compound. Sanger and McCormick were pushing him hard for results, and he assured them again and again that progesterone was the answer, even though it had only worked in about 85 percent of the women tested so far. He was like Orville and Wilbur Wright after their initial flight: confident he had the right idea even though the first try had crashed. As he explained in a letter to Sanger, "[I]f ovulation may be inhibited in a good proportion it should be possible to develop a method for inhibition in 100%."

He didn't offer any details, however, as to why that should be true. All he said was that he intended to find out.

TWENTY

As Easy as Aspirin

ON FEBRUARY 1, 1955, a cold and cloudy Tuesday, Gregory Pincus set out from Shrewsbury to visit Katharine McCormick in Boston. As he drove, snow began to fall, lightly at first and then more heavily.

Pincus had recently returned from Puerto Rico, and he was eager to tell McCormick about his trip. He was also eager, once again, to ask her for money. The snow began to dust the hillsides and coat the blacktop as he rumbled east along Route 9. A car slid in front of Pincus's, and as he tried to avoid a collision he lost control of his vehicle and skated off the road. He climbed out from behind the wheel, shaken but not hurt. He managed to get the car to a mechanic's shop, and from there he hitched a ride to Boston.

McCormick's four-story Back Bay home was like her wardrobe: elegant yet frozen in time circa 1920. A butler greeted Pincus and a maid stood by to fetch drinks. McCormick wanted to hear everything, beginning with the story of his automobile accident. No detail was too small for her attention. She wanted to know if Pincus still believed in progesterone and the assorted progestins he'd been trying. If he were to lose faith, she didn't know what she would do. Pincus said much remained to be done. He still didn't understand exactly how the progestins worked, and he was concerned that about 15 percent of patients still ovulated, even when taking high doses, but he was confident that all the answers would come to him with more time and work.

Late in 1954, Pincus had begun experimenting on animals with a new group of progestins that were many times more powerful than natural progesterone. Two of the compounds seemed especially promising. One, called norethindrone, had been developed by Carl Djerassi of Syntex, the Mexican-based drug company. The other, named norethynodrel, had been developed by Frank Colton of Searle. Pincus had considered a third compound made by the Pfizer company, but Pfizer's owners were Catholic and they refused to supply the chemical when they learned why Pincus wanted it. That left two choices, norethindrone and norethynodrel—two chemical compounds that appeared almost identical in structure. But in testing them on lab animals, Pincus and Chang observed one small difference: Djerassi's compound, norethindrone, caused some of the female animals to develop slightly masculine characteristics. For reasons no one could explain, Colton's formula did not have the same effect.

Pincus informed Searle that he liked norethynodrel, which Searle referred to in its catalogues as SC-4642, and intended to try it as an oral contraceptive for women, possibly on a large scale. He encouraged the drug company to provide the drug to other researchers so they might begin to experiment with it as well. But Searle officials were leery, saying they did not yet understand how or why it worked. Was the pill preventing ovulation, stopping fertilization, or impeding implantation? Was it doing all three? No one knew.

Then there had been another problem, one that Pincus declined to tell McCormick about. Sometime late in 1954, one of John Rock's patients developed a sterile abscess formed by the body's failure to absorb an injected drug, and Al Raymond of Searle wrote Pincus to say the news "throws grave doubt on any interest we might have in this product." If such a side effect were to occur even one time in one hundred thousand, Raymond continued, Searle "would have no interest in promoting the product whatsoever." Raymond said he intended in the future to make certain the Searle label was not attached to the experimental drugs sent to Pincus. The company did not want to be associated with anything that might prove remotely dangerous. "[W]e will send it to you unlabeled," Raymond wrote, "and you can label it as you see fit."

One little abscess, however, didn't worry Pincus much. One thing he hadn't mentioned to officials at Searle or even in his letters to Sanger and McCormick was this reassuring fact: Scientists working at the Worcester Foundation had started giving the progestins to their wives, sometimes for contraception and sometimes simply to control their periods. The wives, said Anne Merrill, who worked as a lab assistant at the Foundation in the 1950s, "didn't want to be bothered with menstruals when they were traveling."

Pincus knew as well as any scientist in the country the many ways a research project might collapse. Funding might vanish. A competitor might win the race to the finish line. Bad publicity might scare away test subjects or bring down the wrath of the Catholic Church or government. Drug companies might lose interest. Results might prove inconclusive or worse. But as he sat in McCormick's lavishly decorated home, with temperatures outside dropping quickly but no more snow falling, Pincus didn't dwell on the many ways the project might fail. Instead, he told McCormick about why he still liked Puerto Rico for trials. A successful birth-control program might become a prototype for other nations. The women were familiar with and often eager for birth control. Many of the doctors and nurses on the island were trained in America and spoke English well. It was true that doctors in Puerto Rico and in the United States initially had had difficulty finding volunteers for the tests, but Pincus settled on a new plan. This time he would begin with nurses and medical students. He would drop the endometrial biopsies, which were extremely uncomfortable and scared off participants. In addition, the students would be required by the faculty to participate as part of their studies. If the young women were worried about the stigma of being birth-control subjects, Pincus had a solution to that problem, too: he would label the project a study of the physiology of progesterone in women. And once the nurses and students enrolled, word would spread that the substances were safe and effective. From there, it would get easier.

Pincus told McCormick the new drugs would cost about fifty cents a gram, which would mean expenses of about five thousand dollars for the treatment of one hundred women during the first year of testing. In addition, he would need money for doctors, nurses, secretaries, travel, and printed materials. For the first year, the total operating expenses would probably be about ten thousand dollars. Maybe more.

As usual, McCormick assured him that money would not be a problem. She was prepared to pay for the entire operation.

Once, McCormick had devoted her full energy to the care and possible cure of her deranged husband. Now, it was the Pincus research project that occupied her. She had no other cause, no other mission, although it was inevitable that a woman of great wealth would sometimes find herself tangled in business affairs. Indeed, it seemed at times that her financial affairs were all she had. With no dear friends or family to surround her, and only a maid and butler to keep her company at home, she spent much of her time speaking and corresponding with her lawyers and accountants. She worried a great deal about her Swiss chateau, which was too big to sell as a home and too expensive to be purchased by a school. Meanwhile, upkeep was costing her thousands of dollars a year, sucking away time and money she would have preferred to spend on birth-control work.

McCormick continued to make donations to Planned Parenthood, but she preferred dealing directly with Pincus and Rock. She phoned and met with the men often, putting to use the knowledge of biology she had received at MIT, as well as the expertise in hormones she had picked up while searching for new drugs to help her late husband. Pincus and Rock visited her at home. She sometimes hired a stenographer to take notes on the meetings with the scientists so her reports to Sanger would be complete and accurate. Once Pincus sent his daughter to McCormick's home with a progress report. Her home was dark and foreboding, but McCormick spoke openly to the young woman about sex, saying how important she believed it would be to separate copulation from procreation, adding that even sex between women might become more meaningful and acceptable once the birth-control pill caught on. Laura was startled and charmed. When it was time to go, McCormick offered to pay her subway fare home. She summoned the butler, who brought a silver tray loaded with coins. McCormick picked up two dimes and handed them to her guest. Only later did Laura notice that the coins were dated 1929. McCormick had probably been saving them since the start of the Great Depression.

McCormick's passion for the birth-control project was so strong she sometimes offered money without being asked. When she heard, for example, that Rock was retiring from Harvard and would be forced to give up his practice at the hospital, McCormick bought a building across the street from the hospital so Rock could continue seeing patients and conducting experiments. She did not want to lose any time while he made his career transition.

McCormick behaved liked the owner of a fledgling business. She let Sanger do the marketing and trusted Pincus and Rock with the technology, but she supervised, and she wasn't afraid to tell the others what they ought to do.

Now, in the first months of 1955, McCormick grew more confident—confident enough to send Pincus a check for $10,300 to pay for progesterone experiments in Puerto Rico. That was in addition to the $20,000 she'd sent to Planned Parenthood in support of the same project. "I do not want him," she wrote in reference to Pincus, "to be in any way held up on this work for lack of funds."

※

Important pieces were coming together. Pincus believed in the progestins. Rock was conducting experiments on women. Puerto Rico offered a possible proving ground.

Before, McCormick had urged members of the team to be tight-lipped about their work. Now she changed her mind, believing that the time had come to tell the world an important discovery was close at hand. Perhaps she was encouraged by the progress of Jonas Salk and the others who had worked in search of a cure for polio. In April 1955, newspapers around the country carried banner headlines such as this one, from the *Pittsburgh Press*: "POLIO IS CONQUERED." The stories that ran under the headlines told of mothers crying and doctors cheering the news. New York City even offered Salk a ticker-tape parade, an honor he declined.

If Salk had done it, and done it so swiftly, why couldn't Pincus? Americans were waking up to the dangers of the population explosion, and they no longer perceived the problem as one affecting only developing countries. The sense of urgency was not as great as the urgency surrounding polio, but it was real and growing. The U.S. Census Bureau issued a report predicting that the nation's population would reach 221 million by 1975, an increase of about 35 percent. At the same time, farm populations were falling and Americans were moving to the cities and suburbs. It was not difficult to imagine that they might begin to feel crowded, that jobs might become scarce, or even that food supplies might falter during hard times.

In February 1955, James Reston of the *New York Times* wrote a story that had people buzzing. "Since Dwight D. Eisenhower became President of the United States, the population of this country has increased by 5,496,000," Reston wrote. "The total on January 1 of this year was 163,930,000—38,351,237 more than when Herbert Hoover left the White House in 1933." Every day, there were seven thousand more people being born than dying in the United States, Reston's story noted. It wasn't Eisenhower's fault; it was the booming economy. But if the economy slumped and population growth continued at its current rate, he warned, the United States might suffer a steep decline. The Cold War might be lost. America would struggle to care and provide for its teeming masses, especially its poor and elderly, and lose its edge over the Russians.

> *The school shortage, the teacher shortage, the job shortage, the housing shortage, the hospital shortage, the nursing shortage, the power shortage, the shortage of roads, the forthcoming fight of labor for a guaranteed annual wage, the controversy over increased mechanization of industry, the disputes over wage rates, farm income, old age pensions, health insurance, and development of our national resources—all go back to the fact that America has a severe case of growing pains.*

The Baby Boom was in full swing and no one knew where it would lead. The headline writer for the *Times* summed it up nicely: "Babies, Babies, Babies—4,000,000 Problems."

A few months after the *New York Times* story appeared, leaders in the birth-control movement gathered in Puerto Rico for a conference sponsored by Planned Parenthood and designed to stir interest in contraception in Latin America. Sanger was too ill to attend and Pincus was too busy, but Dr. Rice-Wray gave a report that earned yet another mention in the *Times*. "When all Puerto Rican parents can have the number of children they want and can properly care for, much of the misery and desperation of our poorer classes can be eliminated," she told the assembly. "Then employment opportunities, schooling, housing, medical and welfare service will have a chance of meeting the needs of the people." During the conference, Planned Parenthood officials appealed to the World Health Organization to make child-spacing education part of its worldwide program of preventive medicine. They also asked the United Nations to recognize a woman's right to birth control as a basic human freedom. The United Nations (with the United States abstaining from the vote) rejected the proposal.

The Catholic Church had threatened to picket the conference in Puerto Rico, but the pickets never arrived and the conference went off smoothly.

Week after week, population control made more headlines, and with each headline came an increasing sense that the problem was real, that the world's natural and economic resources would never keep up with the extraordinary growth in the number of people inhabiting the earth. But there was also a growing sense, especially in America, that the Baby Boom was taking a psychological and emotional toll on the mothers responsible for raising all those children. It had seemed funny in 1950, when Myrna Loy and Clifton Webb had starred in the movie *Cheaper by the Dozen*, based on the true story of a husband and wife with twelve children. The father, Frank Gilbreth, is an efficiency expert who tests his theories on his children. The trailer for the film called it a "laugh riot."

But within a few years of the movie's release, Americans began to take the matter more seriously. In a special report titled "The Plight of the Young Mother," *Ladies' Home Journal* reported that women were working as much as a hundred hours a week for their families—far more than their husbands—even when the women were in poor health. Was this really the best way to raise children? It was a question, the magazine declared, that "demands national attention."

"We don't have a bathtub and we have small children, so I have to bathe all three children in the kitchen sink," Mrs. Edward B.

McKenzie, a mother of three from St. Louis, told the magazine. "It's a problem to get supper cooked with three hungry, tired little babies, and then fly and get the dishes all done, and fly and put them all three in the sink, and then get the baby to sleep first. Then I read the other two a story, and get them to bed. All that time I am wondering, 'Can I get it all done today?' "

Another woman said she gave herself a break from the hectic pace by not washing the family's clothing one day a week. Another said she found a few moments of peace each day by "getting outdoors" to hang out the wash and then again to bring it in when it dried.

Then there was Mrs. Richard Petry, a mother of four from Levittown, Pennsylvania, who insisted that her husband let her work once or twice a week. Why? "To see some people and talk to people—just to see what is going on in the world," she said. Mrs. Petry found a job at a department store, working between six and nine hours a week, but after three weeks of covering for her over a span of hours that added up to only one day, her husband couldn't take it anymore. "I wouldn't have your job for anything," he told her.

One woman was asked if she'd ever had a vacation from housework. "Just in the hospital, having my babies," she said, adding, "if you call that a vacation."

※

Soon after the conference in Puerto Rico, a science reporter for the United Press Association broke a big story that, while fuzzy on details, got much of the thrust right. It began:

Scientists striving to give the human race a simple and sure way of controlling its prodigious and alarming fertility—for example, something as easy as taking an aspirin—believe they are on the verge of success. They're talking very reluctantly when they talk at all, since no scientist wants to rouse false expectations. But this

> *writer has been given good reason for believing that several easy "aspirin tablet" ways which act on the fertility of animals now are being tested—very quietly and privately—in human beings.*

The United Press story went on to say that the pill would likely use hormones to "antagonize" the body's production of sperm or eggs. Planned Parenthood had already spent about $300,000 on the research, the writer noted, although he either didn't seek or else failed to obtain comment from any of the researchers involved.

As the publicity intensified and newspaper editorials expressed growing concern over population growth and support for the development of better contraception, McCormick decided there was no longer any point in keeping the Pincus project quiet. She and Sanger were receiving little support from Planned Parenthood and none from government. Some within the movement were even beginning to wonder if Sanger was losing her grip. She spent much of her time in bed, and she relied on an ever-tightening circle of wealthy patrons, McCormick foremost among them, to fund her work. Like Pincus at the Worcester Foundation, Sanger appeared to be operating the International Planned Parenthood Federation on a month-to-month, project-by-project basis, raising funds as she went along, with no endowment, no safety net, no long-range plan. If Sanger had died, it's likely the whole operation would have collapsed.

With no time to waste, McCormick asked Pincus if he would consider traveling with Sanger to Tokyo in October to present a paper on his discovery at the Fifth International Conference of the International Planned Parenthood Federation.

Never one to avoid publicity or to say no to his wealthy patron, Pincus agreed.

He finished his visit to McCormick's home and stepped into the frigid Boston air to catch a train back to Worcester, his car still laid up from its losing battle with an icy road. When Pincus left, McCormick sat down and wrote a five-page letter to Sanger, telling her about Pincus's frightening crash and all the exciting things he'd had to say in their three hours together that afternoon.

Yet despite all the encouraging news, McCormick closed with her usual impatience: "I do wish the field tests were not so desperately slow!" she wrote. She mailed the letter special delivery, eager that Sanger read it as soon as possible.

(*Right*) Margaret Sanger, shown here in 1922, became one of the country's first crusaders for contraception and sexual freedom. She longed for a "magic pill" that would separate sex from reproduction. *(Margaret Sanger Papers, Sophia Smith Collection, Smith College, Northampton, Mass.)*

(*Bottom, Left*) In this edition of Sanger's *Birth Control Review*, from 1923, she illustrated with little subtlety her view of the impact of unplanned pregnancies. *(Sophia Smith Collection, Smith College, Northampton, Mass.)*

(*Bottom, Right*) A beautiful beginning: Katharine and Stanley McCormick pose at the site of their wedding in Switzerland, 1904. *(Courtesy MIT Museum)*

BIRTH CONTROL REVIEW

Edited by Margaret Sanger

TWENTY CENTS A COPY NOVEMBER, 1923 TWO DOLLARS A YEAR

Official Organ of
THE AMERICAN BIRTH CONTROL LEAGUE, INC., 104 FIFTH AVENUE, NEW YORK CITY

(Top, Left) In 1925, Sanger and Charles V. Drysdale of London led the Sixth International Neo-Malthusian and Birth Control Conference, where they promoted contraception as a means for checking overpopulation, preventing war, and extending the span of life.

(Top, Right) In 1929, Gregory Pincus appeared poised for a brilliant career at Harvard. *(Courtesy of Laura Bernard)*

(Right) Katharine McCormick not only donated money to her favorite causes but also took to the streets, in this case with the National American Woman Suffrage Association. *(Courtesy MIT Museum)*

(Right) McCormick in 1914. *(Courtesy MIT Museum)*

(Bottom) After World War II, birth control became more socially acceptable, thanks in part to advertisements such as this one from Planned Parenthood suggesting smaller families made the nation stronger. *(Margaret Sanger Papers, Sophia Smith Collection, Smith College, Northampton, Mass.)*

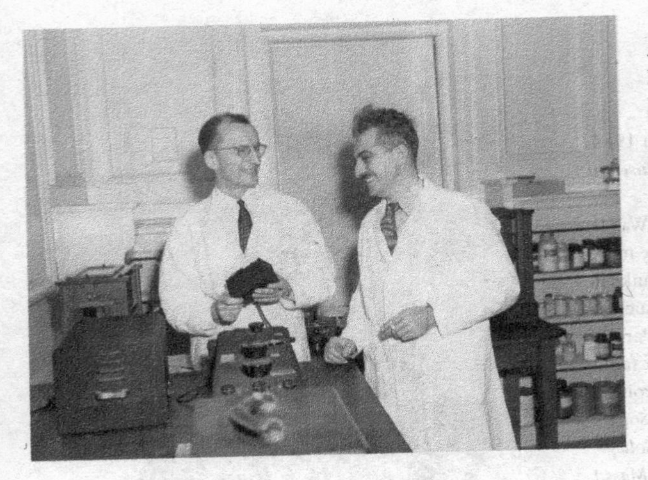

After being dismissed by Harvard, Pincus teamed with Hudson Hoagland, pictured here in 1945, to launch the Worcester Foundation for Experimental Biology. *(University of Massachusetts Medical School Archives, Lamar Soutter Library, University of Massachusetts Medical School, Worcester, Mass.)*

The Worcester Foundation made its home in a converted residence in Shrewsbury, Massachusetts. Pincus had his office in the garage. *(University of Massachusetts Medical School Archives, Lamar Soutter Library, University of Massachusetts Medical School, Worcester, Mass.)*

While living in his laboratory, biologist M. C. Chang conducted some key work in the development of the first oral contraceptive for humans. *(University of Massachusetts Medical School Archives, Lamar Soutter Library, University of Massachusetts Medical School, Worcester, Mass.)*

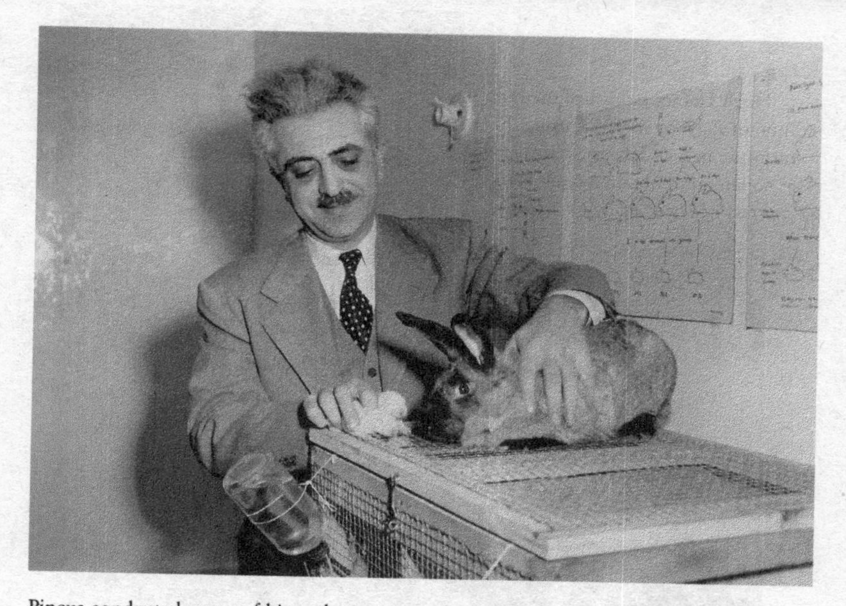

Pincus conducted many of his early progesterone experiments on rabbits at the Worcester Foundation. *(University of Massachusetts Medical School Archives, Lamar Soutter Library, University of Massachusetts Medical School, Worcester, Mass.)*

Lizzie Pincus, Goody's wife, was a brilliant and sharp-tongued woman frustrated at times by the limited career opportunities for women in the 1950s. *(Courtesy of Laura Bernard)*

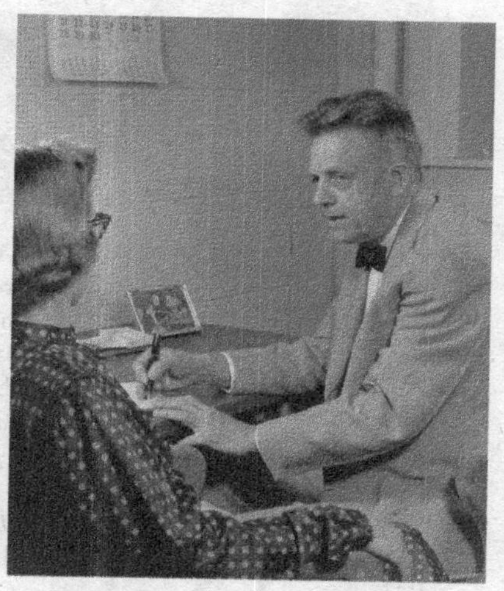

Alfred Kinsey—shown here conducting an interview in 1953—changed American views on sexuality with his large-scale studies on human sexual behavior. *(Courtesy of the Kinsey Institute for Research in Sex, Gender, and Reproduction)*

Hugh Hefner published the first edition of *Playboy* in 1953, using the furniture in his Chicago apartment as collateral to obtain a bank loan. *(Courtesy of The Playboy Enterprises)*

(Above) Large families such as this one—from Tucson, Arizona—were more common before the pill. In 1957, the average American mother had 3.7 children; today she has fewer than two. *(Margaret Sanger Papers, Sophia Smith Collection, Smith College, Northampton, Mass.)*

(Left) John Rock, shown here on a visit to the Kinsey Institute in 1956, was a faithful Catholic but pushed the Church to reconsider its stance on birth control. *(Courtesy of the Kinsey Institute for Research in Sex, Gender, and Reproduction)*

Pincus chose Puerto Rico to test his new pill because the island was poor, crowded, and had a large number of contraceptive clinics, including this one at the El Ejemplo sugar plantation. *(The Harvard Medical Library in the Francis A. Countway Library of Medicine)*

Pincus confers with Drs. John Rock and Celso-Ramón Garcia during clinical trials for the pill, circa 1957. *(University of Massachusetts Medical School Archives, Lamar Soutter Library, University of Massachusetts Medical School, Worcester, Mass.)*

FOR CONTROL OF MENORRHAGIA

ENOVID®

Prolonged or excessive menstrual flow of functional origin can be treated both therapeutically and prophylactically with Enovid.

The supportive action of two tablets of Enovid on the endometrium usually checks abnormal bleeding within six to twelve hours. A daily dosage of one or two tablets is then continued through the intermenstrual interval until day 25 of the cycle. The patient will menstruate approximately three days after discontinuance of therapy.

She is again treated with similar doses from day 5 to day 25 for two or three additional consecutive cycles.

A similar regimen is employed if the patient is seen during the intermenstrual interval. Even though no bleeding is present a dosage of one or two tablets daily is administered until day 25. Therapy is resumed from day 5 to day 25 for two or three successive cycles.

Each tablet of 10 mg. contains 9.85 mg. of norethynodrel, a new synthetic steroid, and 0.15 mg. of ethynylestradiol 3-methyl ether. G. D. Searle & Co., Chicago 80, Illinois.

ENOVID Oral Synthetic Endometropin

(brand of norethynodrel with ethynylestradiol 3-methyl ether)

SEARLE | *Research in the Service of Medicine*

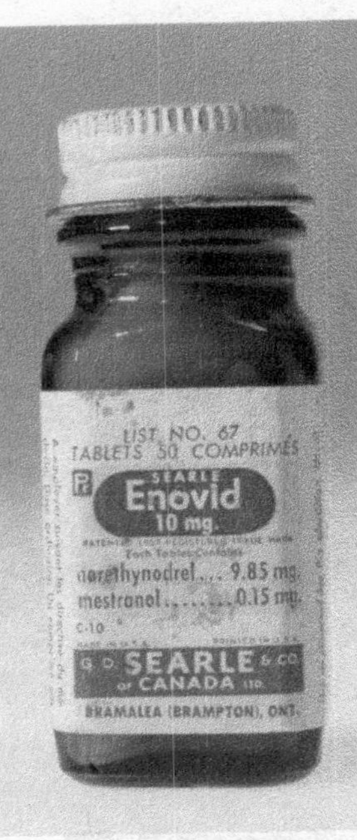

(Top, Left) Before G. D. Searle & Co. won government approval for the sale of a birth-control pill, it marketed Enovid as a cure for irregular menstrual cycles.

(Top, Right) When Enovid was finally approved for birth control in 1960, it quickly became one of the best-selling drugs in the world. *(Photograph courtesy of the Library of Congress)*

(Left) By the time the pill gained popularity, Gregory Pincus was battling cancer. *(Courtesy of Laura Bernard)*

TWENTY-ONE

A Deadline to Meet

NOW PINCUS HAD a deadline: October 28, 1955. On that date, he would stand before a room full of scientists and population control experts and announce that he had done it at last—he had discovered an oral contraceptive that would give women control of their reproductive systems. The fact that he had not yet settled on a precise formula for this pill or tested it on more than a handful of women did not bother him. It was only March. He had seven months.

For Pincus, two progestins—norethynodrel and norethindrone—remained particularly promising, because they were both more powerful than natural progesterone and seemed to work when taken orally. He intended to test both. Beginning in the spring of 1955, Dr. David Tyler recruited twenty-three female medical students at the University of Puerto Rico to serve as Pincus's latest subjects. If the tests went well, there would still be time to include the data in his presentation in Japan. Tyler promised he would do everything possible to deliver results. For starters, he told his female students they were required as part of their coursework to enroll in the clinical trial and if any of them stopped taking the pills and submitting to the urine tests, temperature readings, and Pap smears, he would "hold it against her when considering grades."

Even that kind of strong-arming proved insufficient. Within three months, more than half of Tyler's twenty-three students had dropped out of the trials, either because the pills made them sick or because the tests were so bothersome.

Pincus and Tyler moved on to Plan B. This time, they asked nurses from San Juan City Hospital to enroll in the study. They refused.

On to Plan C. The scientists approached the director of a women's prison at Vega Baja in Puerto Rico to help enroll inmates. The inmates refused, too.

By summer's end, the trials were once again suspended.

Tyler told Pincus he thought he knew what they'd been doing wrong. They'd been counting on doctors, teachers, and prison guards to find subjects, which often meant they were attempting to persuade the uninterested. The trials needed a passionate leader, someone devoted to the birth-control cause, someone who could work on it full-time, and someone who knew how to get out in the community and find the Puerto Rican women who truly wanted and needed a better form of birth control. They needed to find women who wanted what they were offering instead of trying to force it on those who didn't. Otherwise, Tyler said, "it will not succeed."

※

On March 31, 1955, Pincus and his wife arrived in Tucson to visit Margaret Sanger. Goody went almost nowhere without his wife, and his colleagues were well accustomed to her presence at cocktail parties and dinners after scientific conferences. She was the squirt of oil that kept Goody loose, reminding him to get out of the lecture halls and labs and go see the sights. They flew to town on a cool day with thunderstorms sweeping across the desert. Sanger was already playing host to visitors from Japan, so she had no room in her home for additional guests and arranged for the Pincuses to stay at the Arizona Inn, which had been built in 1930 by Arizona's first congresswoman, Isabella Greenway, in part to help create jobs for disabled World War I veterans.

Though the conference in Japan remained seven months away, Sanger was pouring much of her energy into preparations. She vowed to friends and supporters that she would avoid traveling and would reduce her workload so she could build up strength for the big trip. She continued to take the painkiller Demerol, as well as nitroglycerine for her heart, but she'd recently managed to wean herself from Seconal, which she took for insomnia. "I have given up sleeping pills entirely!!" she wrote to a friend. "At first it was terrible, just lying awake & thinking, then reading then writing & finally I thought of warm milk & a gigger of brandy. I was asleep in five minutes. I took milk with less & less brandy & now I do not need anything."

Though Pincus's pill was still almost entirely untested on women and he hadn't even settled on exactly which pill he intended to test, Sanger believed that the biologist's announcement would be the big news from the conference. At the same time, Lader was about to publish his biography of Sanger. With her life story in print and the great goal of that life seemingly within reach, these should have been heady days for her. But since her first heart attack in 1949, her eccentricities had grown more pronounced, and so had her consumption of alcohol and drugs. She had begun gathering her papers and public correspondence so that they might be preserved in the archives of Smith College, but reading through the yellowed pages of her youth opened "veins of sadness," as she described it. The sadness only deepened as old friends and lovers died off one by one. On the advice of her friend, Juliet Rublee, Sanger enrolled in a Rosicrucian mail-order course to help her learn to communicate with the "cosmic forces."

Sanger was becoming increasingly self-absorbed. Her behavior was too erratic to make her effective as a leader of a big organization like Planned Parenthood. But she was still able to focus on smaller, more precisely targeted projects, and the conference in Japan was just such a project.

By the 1950s, Japan's population was about ten times denser than that of the United States. Abortion rates in Japan were so high that the government, hoping to promote better safety, became one of the first in the world to legalize the procedure. In 1951, more than 638,000 legal abortions were performed there. More than twenty thousand midwives worked in Japan, and Sanger was convinced that if the Japanese government supported the introduction of a birth-control pill, the midwives would promote its use and quickly change women's practices. Once the pill came into use, she said, she was confident that the demand for abortions would decline. If it worked in Japan, the same approach might work throughout Asia, where the looming population crisis was most pronounced.

In Tucson, Pincus and Sanger huddled over the program and schedule of presentations for the conference, with Pincus suggesting "more scientific" titles for some of the sessions. Sanger hosted a dinner party for the Pincuses, inviting the president of the University of Arizona among other college officials. She also invited a group of young women—"and their Gynecologist"—to a cocktail party. When Pincus told the young women about his progesterone experiments, several of them volunteered to be his subjects, saying "they would like very much to be progesterone guinea pigs."

After the party, Sanger wrote to McCormick, sharing her impressions of the Pincuses. "I hope to tell you how important it is for his wife to be with him," Sanger wrote, "especially to guard against his Bohemian apparel. She is quite a person in her own, and is evidently a very necessary help to him."

Like McCormick, Sanger trusted Pincus. When he told them not to worry about the clinical trials, they believed him. Then again, they had little choice.

※

By the fall of 1955, John Rock wrote to I. C. Winter at Searle with the preliminary results of his tests with the new progesterone compound, known among the scientists as norethynodrel, or SC-4642.

"It looks pretty good," he wrote.

At that point, he had only tested it on four women.

When Rock told Searle he thought it would be a good idea to expand the experiment by providing the progestin to other researchers, Winter, the company's clinical research director, said he wasn't sure he could encourage scientists to try a compound when he didn't understand how or why it worked. Rock's tests showed that progestins stopped the pituitary gland from producing the hormones that signaled the ovaries to release eggs. But the pill had other effects that were not yet clearly understood. It also appeared to change the consistency of the cervical mucus, for example, making the mucus more hostile to sperm. Progesterone also seemed to make the endometrium less hospitable to eggs. Were all of these effects combining to prevent pregnancy, or would any one of them have worked alone? And what if the progesterone compounds had other effects? What if it stopped the production of cortisone? What if the progestins did some damage to the ovaries that Rock's tests had not yet revealed? What if there were long-term effects no one could imagine?

Rock said he was relatively confident that the Searle progestin was safe and that it wouldn't affect a woman's ability to get pregnant. He was optimistic enough to encourage Searle to promote the drug more widely, but he was nowhere near as optimistic as Pincus. He wanted more tests, and he wanted to publish the results of those tests in a respected journal before making any announcements. If the pill proved flawed in any way, the publicity would backfire. Women would become wary. Testing would become even more difficult. The Catholic Church would harden in its opposition. For all those reasons and more, Rock urged Pincus not to go to Japan.

※

As the conference approached, Pincus made no progress in Puerto Rico. Testing there was at a standstill. Back in Boston, Rock continued to plug away with his small group of patients. Though the size of the study was ridiculously small, the results were good. Both norethynodrel and norethindrone—the former made by Searle, the latter by Syntex—appeared effective. Best of all, they worked at doses of only ten milligrams per day, which was one-thirtieth of the progesterone dose Pincus and Rock had been giving earlier. Perhaps losing faith in Puerto Rico, Pincus told McCormick that he hoped during his visit to Japan he might find doctors willing to start clinical trials in and around Tokyo.

Pincus was always respectful and appreciative in his letters to McCormick, but he was also often vague. He mentioned "a fair drop off" in the number of women participating in trials in Puerto Rico, but he didn't tell her that the drop off was from twenty-three to ten. Nor did he let on that even among the ten women who participated, some had done so haphazardly, failing to take all their pills and failing to submit to all their tests. Nor did he mention that some of the lab specimens couldn't be tested because they'd been accidentally overheated during transport between San Juan and Shrewsbury. In a letter to Dr. Tyler, Pincus confessed that "there is very little [data] worth reporting," but he was not so frank with his biggest sponsor. He neglected to mention that he was coming up empty in his attempts to recruit more women for the study.

As he prepared to depart for Tokyo, Pincus was about to make one of the greatest bluffs in the history of modern science. He was preparing to announce that an oral contraceptive for humans was nearly ready when, in fact, he had not yet decided which form of the contraceptive worked best and at which dose. If that weren't enough, he still hadn't found enough women willing to serve as test subjects.

Some scientists would have been uneasy, but not Pincus, who had the IQ of an Einstein and the nerves of a card shark. He had shown it at Harvard, where he boldly pronounced successes that might have benefited from more careful review. He'd shown it when he'd seduced Lizzie by telling her he was a sexologist. And he'd shown it again when he rebuilt a career that had been in tatters, launching the Worcester Foundation and taking charge of the Laurentian Hormone Conference. Now, in his pursuit of a birth-control pill that might have enormous social and economic impact, he was relying not only on that bravado but also on cunning.

Even the people working most closely with him did not know some of the crafty calculations Pincus had made. McCormick, for example, had no idea that Pincus was drawing pay and receiving stock from G. D. Searle & Company. When Pincus told McCormick that Searle had agreed to supply drugs for testing at no charge, he didn't let on that they might have had a financial interest in doing so. He also didn't mention—although McCormick might have noticed if she checked his expense reports—that when Goody and Lizzie traveled as part of their research work, Lizzie went on shopping sprees and charged her purchases to McCormick.

It's possible McCormick knew about the travel expenses and didn't care, just as she didn't care when she agreed to pay for the renovation of a motel to provide housing for visiting researchers in Worcester; Pincus hired his wife as the decorator and his wife purchased the furniture from an uncle in Montreal. It was not unusual in the 1950s for scientists to accept gifts from drug companies or to let those same companies sponsor their travel. Pincus's behavior was not far from the norm. But it was also true that McCormick was generous and tolerant. Money meant little to her. She had no children to inherit her wealth. She did not collect art or amass real estate. The Pincus project was her great passion. If Mrs. Pincus wanted to buy paintings or pearls, McCormick was not going to raise a fuss about it.

Pincus planned to take not only Lizzie with him to Japan, but also his daughter Laura and one of Laura's college roommates. The two young women would serve as assistants to Sanger, arriving well before the start of the conference to help take care of logistics. Goody and Lizzie would spend a few days in Los Angeles, followed by a few more in San Francisco, and a few more after that in Hawaii before reaching Japan on October 15. After Tokyo, they planned to visit Hong Kong, Bombay, and New Delhi, where they would meet with researchers, doctors, and political activists concerned with birth control.

Before leaving on his journey, Pincus received a $10,000 check (the equivalent of about $87,000 today) from McCormick's personal account to pay for his continued work in Puerto Rico and his upcoming travels. As always, she was prepared to spend whatever the job required, and as always, she wanted the action to unfold quickly. When John Rock asked if he might accompany Pincus to Japan, perhaps because Rock hoped to keep Pincus from making too many promises, McCormick said no. It wasn't that she didn't want to spend the money; it was that she couldn't bear the thought of having both men stop work on the pill.

※

Despite her health problems and despite the seemingly endless delays in testing the new pill formulas, Sanger was once more filled with hope. The publication of Lader's biography had given her a new round of press attention. It had been forty years since she'd fled to England to escape a federal indictment. Now, at seventy-five, she was seen not merely as a relic but as a legendary if somewhat wobbly crusader. Women had been more tenacious when she herself was a young woman, she told a young female reporter for the United Press syndicate. Now, the rebellious spirit seemed to have died in many of them. "You talk to young college women now and they say there isn't anything to do," Sanger complained. The reporter asked her to name the cause she would champion if she were young. Assuming that the birth-control problem had been solved, she said, she would battle to improve the conditions of women in prison. She encouraged women to find the issue that mattered most and fight for it. "You must have faith in it," she said. "I still have it."

Sanger had faith, but it was still not clear if she would live to see it rewarded. With less than three months to go before the conference in Japan, sharp pains radiated from her chest and she checked into Cedars of Lebanon Hospital in Los Angeles, afraid she was having yet another heart attack. After three days of observation, doctors said it had probably not been a heart attack but angina, chest pain caused by heart muscles that don't get enough oxygen-rich blood. It was another sign of her advanced coronary heart disease, but it was not serious enough to keep her in the hospital. In a concession to her health, she told her family she planned to retire as president of the International Planned Parenthood Federation—but only after she attended the conference in Japan.

TWENTY-TWO

"The Miracle Tablet Maybe"

On a cool October day in 1955, Margaret Sanger arrived in Japan. Her supporters waved Japanese and American flags. A man cradled a big bouquet of flowers. Newspaper reporters crowded around, their pens ready, and photographers aimed, clicked, and wound their film to snap again, capturing her every move as she stepped off the *President Cleveland* and onto the dock at Yokohama.

Not many Americans received heroes' welcomes in Japan in the 1950s, but Sanger did.

Only ten years earlier, American bombers had reduced nearly half of Yokohama to rubble and killed more than seven thousand people in a single morning of incendiary air raids. Tokyo had been devastated during World War II, which claimed some two million Japanese soldiers and as many as one million civilians (compared to about four hundred thousand U.S. soldiers and two thousand U.S. civilians). When Japan surrendered, the United States and its allies took complete control of the country. Under the messianic leadership of General Douglas MacArthur, the country was remade. A new constitution stripped power from the emperor and gave it to the people. Business conglomerates were broken up and the economy based on a free-market capitalist system. The Americans were like Christian missionaries bringing their way of life to the pagans, as the historian John W. Dower wrote in *Embracing Defeat*: "The occupation of Japan was the last immodest exercise in the colonial conceit known as 'the white man's burden.' " Instead of rebelling, the Japanese, exhausted and diminished by war, embraced the chance to start over. Defeat had been devastating, but rejuvenation was an extraordinary opportunity that the Japanese pursued energetically. Citizens spoke out in community meetings. Bureaucrats pushed for serious reforms. New celebrities shot to fame. New religions sprang to life. The country was in chaos, but it was a thrilling kind of chaos, and when Margaret Sanger arrived in 1955, she was seen among the Japanese as the greatest kind of American hero, perhaps because her story paralleled their own in certain ways. She had fought the American government. She had suffered because of the American government. But she had stood up for her beliefs even as she put her faith in the system, trusting that democracy and free speech would let her message be heard. Even the losers got a chance in democracy.

It helped, too, that Sanger had visited Japan several times before the war and expressed her admiration for the culture and its people. On her first visit, in 1922, she became a media sensation, followed by reporters and photographers everywhere she went, her words and ideas carried in hundreds of published articles. She had been young and energetic then, and she had offered Japanese women hope for a better future. Throughout much of Japanese history, women had been regarded as the property of their fathers and husbands. They were subordinates—courtesans, prostitutes, military pawns, and servant-wives. Women trying to control their fertility were forced to rely primarily on abortion, which they justified by saying that they were returning their unborn children to the gods. By 1922, Japanese prostitutes were still advertised as if they were dinner entrees, with prices printed on cardboard menus, including rates by the hour or night, and girls as young as ten still worked thirteen-hour shifts in silk-spinning factories. But attitudes began to change in the 1920s, which is why Sanger was not only welcomed but cheered. Government grew more permissive and reform movements took hold. Overpopulation emerged as a social issue, too.

Japan was one of the most densely populated places on earth. Already, the country couldn't grow enough rice to feed its people and was forced to rely on imports. Though her speeches were heavily censored, Sanger gave more than a dozen lectures and met hundreds of women. She became a symbol of strength and independence that would inspire Japanese women for years to come.

She inspired local activists to begin speaking out, including Baroness Shidzue Ishimoto, who said that Sanger had appeared "like a comet" and left "a vivid and long-enduring impression." Ishimoto would go on to model her own career on that of Sanger's, arguing that birth control was the tool women would use to build independent lives. Contraception gained wider acceptance in the years after Sanger's first visit. Planned Parenthood pamphlets were translated into Japanese and distributed by doctors and Buddhist ministers. Companies making contraceptive devices used Sanger's name and photograph (without permission) in their advertisements, and some even named their products after her. There were "Sangerm" spermicidal suppositories and "Sangai" diaphragm-and-jelly kits. Of greatest concern to the crusading feminist was an illegal abortion-inducing medicine simply called "Sanger" that was falsely advertised as having been "improved" by Margaret Sanger herself.

Sanger loved Japan, not only because the people there treated her like an idol but also, she said, because she saw "none of the ranting bitterness" that she encountered in her own country, "no priests denouncing me as an advocate of unbridled sex lust, no celibate clergy assailing me as the arch-apostle of immorality."

The new constitution in Japan had given women the right to vote and organize labor unions. Sanger's protégé, Ishimoto (now remarried and known as Shidzue Kato) won election to the Japanese Diet in 1946. But even as women were gaining more rights and power, abortion rates in the country rose sharply because of high unemployment rates and housing shortages. The number of reported abortions increased from about 246,000 in 1949 to about 806,000 in 1952, and the number of sterilization operations would jump from 6,000 in 1949 to more than 44,000 in 1956.

Sanger, who opposed abortion as a method of birth control, believed that Japan needed an oral contraceptive more desperately than most nations. She also believed the country was well positioned to take advantage of innovations in contraception. Literacy rates were high. Midwives were active even in remote villages. Word of the new method would spread quickly. If it worked in Japan, Sanger strongly hoped it might work all over Asia and all over the world. Women would gain control of their bodies, gain control of family size, gain control of government. Before long there would be no more war. Independent women would lead the world through an era of unprecedented peace and prosperity. That was the dream, anyway. But it wouldn't happen with IUDs or condoms. Only Pincus's new and improved product would do the job.

By the time of her arrival, Sanger had already made Japan aware of birth control's urgency. Her goal for this visit was to announce to the world that a birth-control pill was close at hand and to get scientists in Japan and other countries to begin spreading the news and participating in clinical trials. If large-scale trials in Puerto Rico failed, Sanger hoped that Japan, where she seemed to have the golden touch, might be another option.

Ten days before Pincus's scheduled speech, she began teasing reporters: When the conference opened, she said, an American scientist would announce that he had nearly perfected an inexpensive, all natural, oral contraceptive that could be eaten like candy. Within a year the drug would be available all over the world, she said in a remark that qualified either as wishful thinking or willful misinformation. "Within a year it will be cheap enough so it can reach the very poorest of people," she told reporters. "The pill will eliminate contraceptive devices."

As great as this candy-like contraceptive would be, she boasted, Dr. Pincus had something even better in the works: a single injection that would provide a woman with six months of protection from pregnancy.

Now Sanger was getting carried away. Pincus had discussed the possibility of such a contraceptive with his colleagues, but he wasn't working on it.

It didn't matter, though. Japan awaited Pincus's arrival, and the world stood by to hear the details of the pill, which Sanger referred to in one statement as "the miracle tablet maybe."

※

Lizzie and Goody Pincus arrived in Japan on October 15, two weeks after Sanger, and spent most of the week leading up to the conference as tourists. In crisp autumn weather they traveled with other scientists and their wives to see the Kegon Falls, the Chugenji Temple, and the trout hatchery at the Senjogohara Plateau, where Lizzie, cigarette jammed between white-gloved fingers, lectured her Japanese hosts on how best to cook the fish. The Pincuses ate tempura and received lessons in how to use chopsticks, but when geisha girls offered sake, Goody declined. Though normally fond of a strong drink, he complained of stomach trouble and asked for milk instead.

In the days before his big speech, while Lizzie shopped, Goody visited universities and lectured graduate students. They stayed at the Frank Lloyd Wright–designed Imperial Hotel, one of the few structures in the city to survive both the earthquake of 1923 and the American firebombing of Tokyo in March 1945.

※

The conference began at 9 a.m. on Monday, October 24. Four hundred men and women—including dozens of foreign scientists—crowded the auditorium of the Masonic Building in central Tokyo. A great wave of applause greeted Sanger as she stepped to the podium in flat shoes, looking like an elderly churchwoman with a dowdy sweater over a dark blouse and skirt. A pillbox hat sat lightly atop her wavy, gray hair. A translator stood by her side.

As she gazed out at the crowd, Sanger smiled and said she felt as if she were at home in Japan. As much as she loved the country and its people, though, she said it was clear as she walked the streets of Tokyo in recent days that the place was growing too crowded. The problem was not Japan's alone. It was a global one, and solving it would require more than education. It would require more than teaching women to be more assertive and teaching men to control their libidos. That is why she had come to Japan yet again, to suggest a better way.

"This conference is bigger than the previous ones," she said. "This is going to be a landmark event for family planning because we are going to talk about birth-control research, which we were not ready to talk about in the previous conferences."

Pincus's remarks were scheduled for the afternoon of October 28, the fifth and penultimate day of the conference. In addition to the press that would be covering the speech and the birth-control advocates who would be there to help spread the news, some of the world's leading experts in reproductive biology would be in attendance, including the zoologist Solly Zuckerman, who would be knighted in 1956 for his service to England during the war, when he studied the human and economic effects of bombing raids and helped the Royal Air Force choose its targets in the buildup to D-Day. Zuckerman was one of the great polymaths of his time, a friend to George and Ira Gershwin and an expert on the social life of monkeys and apes. Also on hand was Dr. Alan Sterling Parkes, one of England's pioneers in hormone research and already a legend in his field. Like Zuckerman, he would soon be knighted for his contributions to science. The day before Pincus's scheduled remarks, the eminent Dr. Parkes expressed low expectations when he presented a review of various birth-control research projects underway around the world and said that, in his opinion, they offered "little or no hope of early development towards practical application." In theory, Dr. Parkes said, it was possible that a safe compound might be discovered that would prevent the pituitary gland's hormones from reaching their target in the ovaries, but only in theory. "I need hardly say," he continued, "that no such substance is yet known."

Hours before Pincus was to make his speech, M. C. Chang presented one of his own. Though he remained Pincus's close associate, Chang thought it was too soon for his boss to declare victory in the search for a birth-control pill. Like Rock, Chang wanted more time to prove it really worked. He also had a broader concern that a daily pill to inhibit ovulation might be the wrong approach. Even if the price of the pill came down dramatically, he wondered if it would ever be cheap and convenient enough to prove effective for women in the poorest communities. He had a nagging feeling that by creating a pill that required swallowing twenty-one doses a month, drug companies would be the biggest winners. They'd have women hooked for years on their daily pills. Pincus tried to allay his fears. Drug company profits, he said, were a "necessary evil." The pill might not be perfect, but it would do powerful work.

Chang never completely bought it, as his remarks in Tokyo made clear. "Unless and until we know more about the basic mechanism of fertilization or reproductive physiology," he said in his heavily accented English, "to devise an effective measure for its control is only a hit and miss affair."

For almost a month Sanger had been promising a bold new form of birth control, but so far attendees had heard that no such substance was on the horizon. The audience awaiting Pincus's remarks might have been justifiably confused. At the least, they were surely curious as to what he would say.

Was there a miracle tablet or not?

Pincus showed no anxiety as he stepped to the podium. "He was the most supremely confident and self-assured person I have ever met," said his daughter Laura. "Nothing daunted him because he always knew he would succeed."

Pincus began by repeating some of what Chang had said, that synthesized progestational hormones worked effectively in preventing pregnancy in lab animals. He named many of the substances and compared their effectiveness. Norethindrone and norethynodrel were the most promising, he said, and went on to offer details from the studies on humans being performed by John Rock in Boston and others in Puerto Rico. He did not dwell on the small numbers of women involved in those studies, but he made it clear that the findings were preliminary. More testing needed to be done and soon would be. He stressed that the compounds tested thus far produced no harmful side effects in animals, a fact that gave him great hope that the same would hold true in humans. He spoke confidently, his eyes wide under those bushy eyebrows, his hands working the air. The non-scientists in the room might not have understood everything he said, but at least he said it confidently.

He went on:

> *We cannot on the basis of our observations thus far designate the ideal antifertility agent, nor the ideal mode of administration. But a foundation has been laid for the useful exploitation of the problem on an objective basis. . . . The delicately balanced sequential processes involved in normal mammalian reproduction are clearly attackable. Our objective is to disrupt them in such a way that no physiological cost to the organism is involved. That objective will undoubtedly be attained by careful scientific investigation.*

No hats were tossed in the air. No standing ovation greeted Pincus, only polite applause.

Had the presence of men such as Solly Zuckerman and Alan Parkes prompted him to avoid making unnecessarily bold claims? Had he decided to heed John Rock's words of caution? Or was he simply doing what good scientists do: presenting data and letting it speak for itself?

The big headlines Sanger had promised never materialized. Pincus's remarks were greeted with near silence in the press and even a note of skepticism among his peers.

"Promising though they may have appeared at first sight," Zuckerman said, "I think it is . . . fair to conclude that the observations reported by Dr. Pincus do not bring us as close as we should like to the goal of our researches."

Zuckerman noted that he had studied the effects of progesterone and estrogen on monkey ovaries in the 1930s. There was nothing groundbreaking in that. The only piece of news, as far as Zuckerman could tell, was that Pincus had put the same hormones in a pill. While a pill could be of enormous help in making birth control more accessible, it wouldn't matter if the substance proved unsafe or unreliable.

"We need better evidence about the occurrence of side effects in human beings," Zuckerman told the gathering in Tokyo. "It is not enough, it seems to me, that we take presumed negative evidence about the lack of side-effects from animal experiments to imply that no undesirable side-effects would occur in human beings. There is an urgent need for prolonged observation before we draw any firm conclusions."

TWENTY-THREE

Hope to the Hopeless

PINCUS LEFT TOKYO and completed his tour of Asia along with Margaret Sanger, his wife, and a few others. It was the first time he'd been exposed for any length of time to Sanger's world—to the rural midwives and doctors and the women they cared for, to the mothers caring for more children than they could afford, to the brothers and sisters sleeping eight to a bed, and to the local and national government leaders who set policy on family planning. Perhaps it reminded him of the work his father had done on the commune in New Jersey, teaching agriculture to the masses, harnessing science to improve the lives of the poor. This was why his work mattered.

A few years later, Pincus would write a letter to a friend who had known his father, saying that his travels in the Far East had helped him think differently about his work. He had begun to realize, he wrote, "how a few precious facts . . . in the laboratory may resonate into the lives of men everywhere, bring order into disorder, hope to the hopeless, life to the dying. That this is the magic and mystery of our time is sometimes grasped and often missed."

He could not control whether the magic of his own work was grasped or missed. His job was merely to explore, expound, and hope for the best. Fortunately for him, though, birth control was becoming part of a broader movement toward social equality and women's rights at the time, even if few people recognized it. And that movement toward equality was helping to make the world more receptive to his work.

By the fall of 1955, humans—and women in particular—were asserting themselves more than ever when it came to controlling their bodies and lives. White, married, middle-class women were nesting in their suburban homes and making and raising children—lots of children—just as the stereotypes of the day said they should. But not all of these aproned suburban housewives were happy about it. Then there were the women who weren't white, married, middle class, or living in the suburbs. They had reasons of their own to be dissatisfied. There were young black women moving from the South to the North and immigrant women arriving from distant countries, all exploring communities that offered new opportunities and new perils. There were smart, young, unmarried women competing with men for spots in law school and medical school. The black women from the Deep South, the immigrant women, and the college women considering careers outside the home had something in common: they recognized that the pursuit of opportunity required independence, and achieving that independence meant avoiding—or at least postponing—motherhood.

In the 1950s, women were voting in roughly equal numbers to men for the first time in American history. The radical feminist movement of Margaret Sanger's youth was gone, but other forms of rebellion were taking root. In the South, women like Rosa Parks, Septima Clark, and Ella Baker helped spark the civil rights movement. In factory towns and in cities, women became union activists. When they married or when they had children and wished not to have more, women turned to doctors, priests, and even newspaper columnists for advice, and they did so without the same degree of shame their mothers would have felt. "Contraception" wasn't a bad word anymore. Even Catholic women were exploring birth control, justifying it in their minds by thinking this was perhaps one area where they knew better than the Church what was right and moral.

When one newspaper advice columnist in Oakland published a letter from a reader who favored birth control, a heated debate unfolded in the paper's pages.

"Why, I know someone else who was born of a poor mother, and one day someone hung Him on a cross and He became the Savior of the world, and others became doctors and nurses, and teachers and poets, and lawyers and truck drivers, and presidents and singers—and some of the best darn people you or anyone else would want to meet," wrote a woman calling herself "Just Darn Mad." Other writers cited their religious beliefs: "A person who frustrates the very purpose and actual basis of the marriage relationship and yet takes the pleasure is cheating God," wrote one eighteen-year-old married woman who was pregnant with her first child. "He attached the pleasure as first an inducement and secondly as a reward, although that seems a poor choice as a reward." God attached pleasure to eating, too, the young woman wrote, which means that if a woman wanted to keep her figure, she needed to watch what she ate. The same went for sex; if a couple wanted a small family, she concluded, they "must curb THEIR appetite!" She signed her letter "Happy Expectant Mother."

A few weeks later, a woman signing her letter as "A Practical Parent" wrote to the same newspaper columnist to say that she wished "Happy Expectant Mother" well but wondered if she would be quite so cheery after her third or fourth child arrived. "Someone ought to inform this young lady," she wrote, "that if God intended you to keep producing babies year after year, He wouldn't have made it so easy to avoid it."

Another woman wrote to say that she had begun using birth control when her first pregnancy ended in miscarriage and her doctor told her that another pregnancy might prove fatal to her and her baby if she didn't wait at least two years to recover. "Can anyone say I'm a sinner because I selfishly want to live and be able to bear healthy, normal children?" she asked. "I don't believe so. . . . We are expecting a baby in March. . . . I wish people would remember there are two sides to every story. . . . We should be tolerant of all religions and beliefs." She signed her letter "A Very Happy Person."

And those were only the respectable women. Others, like Janis Joplin of Port Arthur, Texas, rebelled more brazenly against the old moral codes. "I wanted something more than bowling alleys and drive-ins," Joplin said, describing the teenaged years before she became a rock star. "I'd've fucked anything, taken anything."

Women such as Joplin were searching for lives radically different from their mothers'. Movies of the 1950s goaded them, making middle-class homes look like prisons and parents look like losers. In *Rebel Without a Cause*, James Dean did in fact have a cause: he was fighting his parents. Many girls were ambivalent if not completely frightened about a future that appeared at times to offer nothing but marriage and children. The writer Marge Piercy recalled, "All that could be imagined was wriggling through the cracks, surviving in the unguarded interstices. There was no support for opting out of the rat race or domesticity. . . . Marry or die!"

Margaret Sanger's recent promises had created an impression that a birth-control pill was close at hand, but young women like Piercy and Joplin were not waiting for something magical to give them their liberation. Neither were they waiting for a pill to let them explore their sexuality. In 1956, Grace Metalious published *Peyton Place*, a novel filled with scenes of rape and incest billed as a story that "lifts the lid off a small New England town." Mothers hid the books under their mattresses. Their teenaged daughters would find them and tear through "the good parts." In one memorable scene, town harlot Betty Anderson is furious that bad boy Rodney Harrington has taken Allison MacKenzie to the school dance. Betty gets Rodney riled up in his car, asks him if he's "good and hard," and then, when Rod is so excited he can hardly speak, Betty jackknifes her knees, pushes him away, and gets out of the car. She tells him to take his erection and "shove it into Allison MacKenzie . . . and get rid of it with her!"

Critics denounced the book as filthy, sordid, and cheap, a likely corruptor of youth. Libraries banned it. So did Canada. Of course, the critics and censors only stoked more interest, and *Peyton Place* became a mammoth blockbuster, sitting atop the *New York Times* bestseller list for fifty-nine weeks. By the end of its first year in print, one in twenty-nine Americans had purchased the book.

"I was living in the Midwest during the 1950s," said Emily Toth, Metalious's biographer, "and I can tell you it was boring. Elvis Presley and *Peyton Place* were the only two things in that decade that gave you hope there was something going on out there."

In 1956, a woman still had to be shockingly bold to admit in public that she liked sex, especially if she was unmarried. Doctors still referred to sex as "the sex act," which, like the preparation of dinner and the ironing and folding of laundry, was considered part of a married woman's household responsibilities. She performed the sex act to make her husband happy or to propagate the species; she was not supposed to enjoy it. Indeed, women who craved sex too strongly were sometimes deemed in need of medical or psychiatric intervention. "Characters like these belong in an asylum," one *Peyton Place* reviewer wrote, "and, as a security measure, the town would be declared out-of-bounds by all civilized people."

Clearly, *Peyton Place* had struck a nerve. A great social change was underway. Everybody had the fever—"a feelin' that's so hard to bear," as Little Willie John sang in the hit rhythm-and-blues song of 1956.

※

It was ironic that the great American sexual burning came as Sanger had all but completely remade her crusade. A mission that had originally been built around the joy of sex and the desire for more of it was now constructed around respectable themes such as population control and sound parenthood. If that approach tended to induce yawns rather than gasps, Planned Parenthood wanted it that way, and Sanger had gone along grudgingly. It wasn't that Sanger or Planned Parenthood had lost interest in sex. On the contrary, the Planned Parenthood Federation of America was one of the only organizations in the world that made a woman's sexual fulfillment part of its official program, offering sex counseling that was often disguised as marriage counseling and working with doctors, social workers, and mental health services to promote sex education.

Sanger was able to get grants from wealthy friends and arrange meetings with world leaders now in large part because she had taken the birth-control movement mainstream, arguing that contraception was a tool for economic growth and political stability. By 1956, that transformation was nearly complete, but Sanger and the organization's other leaders believed there was one important faction they needed to win over before birth control would truly gain widespread acceptance.

The message was summed up in a memo from 1954 that continued to circulate in 1955 and 1956 among leaders of the International Planned Parenthood Federation. It mentioned rumors that the Catholic Church might soon be pressured by its followers "to accept some of the new forms for the rational control of fertility, if and when they are developed." The report continued: "The Church is far from monolithic; it contains many different points of view, and the Pope, like many other authoritarians, can only move a certain distance ahead of the different elements who support him." Planned Parenthood had two choices, according to the memo, which was sent to Sanger among other leaders: The organization could harass the Church and stir controversy as it had done consistently through the years, or it could "avoid controversy, trying to show them the effects of world population . . . and to work with those elements in the Church who want to see a change." Catholics comprised a quarter of the American population. Winning them over, or at least gaining some allies within their ranks, would be a coup for Planned Parenthood.

The Vatican could not have been any clearer at the time in its attitude toward birth control. In 1951, Pope Pius XII spoke to the Italian Catholic Society of Midwives and reaffirmed what his predecessor, Pius XI, had said on the subject: that Catholicism would not sanction any attempt to impede the creation of life during the conjugal act. "This precept is as valid today as it was yesterday," the pope declared, "and it will be the same tomorrow and always, because it does not imply a precept of the human law, but is the expression of a law which is natural and divine."

Anyone disobeying such a strong and clear declaration, theologians said, would be sinning against the faith. But, as John T. Noonan wrote in his definitive account of contraception and Catholicism, the pope did sanction the rhythm method, which he deemed "natural" because it didn't kill sperm the way a spermicide did; it didn't impede the normal processes of creation in the manner of a diaphragm; and it didn't mutilate the body's organs as sterilization did. That led some theologians to contemplate other loopholes, including the one presented by the research of Pincus and Rock: What if a drug allowed women to control or extend their safe periods? And what if this drug were built with or based on ingredients occurring naturally in the body? Would that be natural? Would that be enough like the rhythm method? Would that satisfy the pope?

Some thought it might. The human body secreted progesterone during pregnancy to protect an unborn child in the mother's womb. If science made it possible with natural compounds, why shouldn't a woman have the power to prevent a pregnancy that might endanger her health or harm the well-being of her existing offspring? Could she take a drug during the first six months after the birth of a child, while she was nursing, to make sure she would not get pregnant again?

Wouldn't that be useful? And wouldn't that be just as morally acceptable as the rhythm method?

It was a question neither the Church nor anyone else had been forced to answer, but one the Vatican would soon confront. Not only were Pincus, Rock, and Planned Parenthood pressing the issue, Catholic women were, too. American women were having children at never-before-seen rates in the 1950s. While the average American woman in 1957 would give birth to a record-setting 3.7 children over the course of her lifetime, for Catholic women the average was about 20 percent higher.

Church officials lectured the faithful not to give in to the temptation to use birth control. "You need to think about heaven and hell," one cleric wrote. "You need to think about death, and how God can call you in the very midst of your planning for a comfortable future on earth."

But even worshipful Catholics were conflicted. When a *Catholic Digest* survey in 1952 found that more than half of all Catholics did not regard "mechanical birth control" as inherently sinful, Paul Bussard, the priest who ran the magazine, was so disturbed he decided not to publish the results. By 1955, 30 percent of Catholic women admitted using a form of birth control other than abstinence or rhythm. A growing minority within the Church began to overlook its teachings on this subject, and many Catholics stopped regularly attending the sacraments of confession and communion. In letters to the editors of Catholic publications and in discussions with their priests, Catholic women expressed their dissatisfaction.

"I have had seven children within eight years, despite frantic and distressing efforts to follow rhythm," one woman wrote. "It seems unjust that we who have accepted the responsibilities of marriage should have to practice continence."

In response, priests prescribed prayer and the sacraments. The women, feeling as if their religious leaders were out of touch, responded bluntly: With eight children, who had time for the sacraments? Then there were other priests who told the young women in their flock that the Church might be mistaken in this instance, an approach that had equally devastating effects over time.

"I was taught by the Church that if you used birth control the faces of your unborn children would haunt you on your deathbed," recalled Loretta McLaughlin, who grew up in Boston and went on to become a journalist and the biographer of John Rock. "When I was nineteen years old I confronted my priest. I said, 'I don't believe it.' The priest said to me, 'I wouldn't worry about it,' and he walked away." McLaughlin was so infuriated by the priest's "high-handedness" and casual dismissal of her concerns that she never went to confession again and soon stopped attending Mass.

In 1956, Vatican leaders recognized the deepening chasm between the Church's leaders and its followers. They heard about it from priests and read about it in newspapers and magazines. They also knew that a new birth-control pill was in development, with a Catholic doctor involved in its creation. If the pill reached the market, it would compel the Vatican to make a choice—whether to continue to hold the line against contraception or to moderate its position. John Rock was hoping he might persuade the leaders of his faith to choose the latter. In the meantime, he had found a way to solve, at least temporarily and in his own mind, the moral problem of whether he might be violating the doctrines of his faith by testing an oral contraceptive in the heavily Catholic communities of Puerto Rico. While it was true that the Church forbade contraception through chemical means, there was no explicit ban on experimentation. Technically, he wasn't giving pills to prevent conception; he was giving them to find out if they worked.

Once, another Catholic doctor confronted Rock, telling him he was being naive, that the Church would never accept a birth-control pill no matter how well it worked, how closely it resembled the rhythm method, or how hard Rock lobbied for it. Rock, who towered over the younger doctor, stared him down before speaking.

"I can still see Rock standing there," I. C. Winter recalled of the encounter, "his face composed, his eyes riveted . . . and then, in a voice that would congeal your soul, he said, 'Young man, don't you sell *my* church short.'"

TWENTY-FOUR

Trials

IN FEBRUARY 1956, Pincus flew to Puerto Rico to see if he could salvage the trials. The university students and nurses had all quit. Threatening them hadn't worked, and Pincus saw no point in trying to get them back. He needed a new approach.

Arriving in San Juan, he met with Edris Rice-Wray. Rice-Wray then held two posts: She was the medical director of the Family Planning Association and director of the training center for nurses at the Department of Health's public health unit in Rio Piedras (meaning "river of stones"), a poverty-stricken district of San Juan. These dual roles made her an ideal guide to Puerto Rico for Pincus and Rock. She knew her way around the communities where the field trials would be taking place. She knew the doctors and social workers in those communities. What's more, she had pull with local officials who might offer protection if the work stirred controversy. Rice-Wray—or Edie, as her friends called her—also had something in common with Pincus, Rock, Sanger, and McCormick: she was a rebel. She'd given up a comfortable medical practice in Chicago because she believed women should have access to contraception. She believed birth control would help women overcome some of the fundamental inequalities of being women, that it would free them to seek more education, pursue better jobs, and raise healthier and better-educated children. She liked the idea of a scientific form of birth control, and she was excited that Pincus was working on one. But she was nevertheless hesitant to test new drugs on patients.

"I was kind of scared of it at first, really," she told one interviewer. Rice-Wray didn't want to offer her patients a drug that didn't work, and she certainly didn't want to offer them anything that might do harm. But Pincus was so charming and self-assured that Rice-Wray was won over.

A few weeks after Rice-Wray's meeting with Pincus, John Rock flew to Puerto Rico to talk to doctors and nurses there about how the trials were to be run. Rice-Wray was unsure what to make of Rock at first. He was a pipe-smoking Bostonian, the picture of sophistication, and a Catholic, to boot. Yet here he was in San Juan, wearing an ascot or tie even in the tropical heat, standing erect and proud, eager to go to work in the slums.

"I understand you're very Catholic," she said, "and yet you're in favor of birth control."

"Do you want to know what I think?" Rock told her. "I think it's none of the Church's damn business."

Rice-Wray liked him. She told Rock the same thing she'd told Pincus: that they'd been going about their work all wrong in Puerto Rico. They'd been trying to shove their experimental pills down the throats of women who didn't want or need birth control. But Rice-Wray knew where to find women who would be eager to go along with the experiment, women who were desperate for more effective birth control. In Rio Piedras, where ramshackle slums had recently given way to government-funded housing developments, young women were fighting to escape poverty. Rice-Wray felt certain that many of the women she had met in Rio Piedras would be willing to try a new form of birth control. What's more, the new housing developments offered an ideal laboratory. There would be no "wading in the mud" for the doctors and nurses visiting the community, she said, and because so much of the housing was new and highly desirable, the population would remain stable over months of work. Her efforts there so far, mostly funded by Planned Parenthood and Clarence Gamble, had shown that women visiting clinics in Rio Piedras didn't need to be strong-armed. Rice-Wray was certain that if the pill worked, word would spread quickly and demand for the new drug would be strong.

Rice-Wray began by visiting the superintendent of the Rio Piedras housing development, a man who saw firsthand the effects of overpopulation and the burdens it placed on young mothers. He turned over a detailed list of all the community's residents and promised that his staff would help Rice-Wray recruit subjects. Next, she enlisted a nurse named Iris Rodriguez. Rodriguez was a strong, smart, ebullient woman who knew everyone in Rio Piedras by name. She would make a survey of young couples who had children (to establish that the women in the study were indeed fertile) and wanted to have more. They would screen out women who were over the age of forty or sterilized, as well as women who were making plans to leave the community within the year.

Puerto Rican government officials approved the study but asked that Pincus and his team avoid publicity. As they recruited women, Rice-Wray and Rodriguez were careful to explain that this was a private, scientific study, not affiliated with the government. They compared themselves to the Cancer League or National Foundation for Infantile Paralysis, emphasizing that they were doing research to improve health and help parents manage their family's size.

"It was very easy to talk to mothers about the contraceptive program in neighborhood gatherings or in the health centers," Rice-Wray recalled years later in a speech to the Royal Swedish Endocrine Society, "but very difficult to get fathers to attend such meetings." To find out what fathers thought about birth control, Rice-Wray went to the city jail in San Juan and interviewed prisoners, and she came away with the impression that the men, while reluctant to admit it, were every bit as interested as women in limiting family size.

By the end of March 1956, Rodriguez and Rice-Wray had selected a group of 100 women, as a well as a control group of another 125. Though almost all the residents of Rio Piedras were Catholic, Rice-Wray encountered only one woman who said her religion prohibited her from enrolling in the study. The subjects receiving the birth-control pill were told that they were taking an experimental new contraceptive. The women in the control group were told they were part of a survey on family size. In April, the researchers began dispensing pills.

Pincus and Rock had chosen Searle's compound, norethynodrel, over Syntex's norethindrone. They said it was because the Syntex compound produced slight increases in masculine traits among lab animals, but Carl Djerassi, the chemist who had developed norethindrone for Syntex, believed there was another reason: Pincus had long-standing business ties with Searle and owned shares of the company's stock. They began with a dose of ten milligrams per pill, the same amount Rock had given his patients back in Boston. The women were told to begin taking the pills five days after the start of their next period. Then they were supposed to take one pill a day for twenty days before stopping. If they missed a day, they were supposed to take a double dose the following day. The instructions were complicated and this method of birth control completely unfamiliar to the participants, but Rice-Wray said the women were eager to begin. In fact, she told Pincus, they were "crazy to get the pill."

※

As the tests began in San Juan, Pincus invited Rock to present a summary of the first round of human tests at the thirteenth annual Laurentian Hormone Conference in September 1956. A few months earlier, Rock had urged Pincus not to talk about their work in Japan.

He had been afraid of backlash from the Catholic Church. Now, however, he was ready.

The biggest names in hormone research were there, and, inspired by Pincus's leadership style, they were notoriously aggressive. Speakers were not expected merely to read their scientific papers and bask in a round of applause; they were expected to stand on stage for a grilling by their scientific peers. Rock's paper, coauthored with Pincus and Dr. Celso-Ramón Garcia, was modestly titled "Synthetic Progestins in the Normal Human Menstrual Cycle." Most of the paper was dull, even by scientific standards, but one subsection caught the attention of everyone in the room—the section describing the effects of progestins on the ovulation of fifty women tested at Rock's clinics.

Rock approached this subject cautiously. Though the conferees knew that he and Pincus were on the brink of something potentially big, Rock refused to speculate. He showed slide after slide illustrating the effects of progestins on vaginal tissue and ovarian tissue. He explained in details the results of his patients' urine analyses. The word "contraceptive" never escaped his lips. This was as sexy and provocative as the buttoned-up Rock allowed himself to get: "We are led to suspect," he said, "that ovulation has been inhibited in at least a very high proportion of cases."

Finally, he added that seven of his patients became pregnant after treatment. This was important for many reasons, not the least of which being that Rock's patients in the study had come to him seeking help having children. But that wasn't all. Rock also wanted the scientists to know that the progestins did no harm to the eggs or ovaries.

The scientists in the audience pressed him.

"It seems to me we have anti-ovulation!" a voice cried out.

Rock grinned but refused to take the bait.

"I didn't say it, but I allowed them to know," he recalled years later in an interview, that it was indeed the case. He wasn't being modest and he wasn't afraid of angering officials in the Church. In his mind, it was a matter of showing respect for his Church. "I think I was protecting Catholicism as such, rather than myself," he said. "I had a loyalty to the Church, a loyalty which kind of transcends belief. The time was not yet right to flaunt the contraceptive effect."

That did not stop the scientists from asking more questions. Dr. Edward T. Tyler, a gynecologist from Los Angeles, said he recently had been testing norethindrone on his patients with irregular menstrual cycles, but some of his data were different from Rock's and Pincus's. He wanted someone to explain and wasn't getting answers.

"Dr. Pincus and I discussed this until 2 a.m. last night," Dr. Tyler said. Pincus found consistent declines in pregnanediol—an inactive product formed from the breakdown of progesterone—while Tyler found no such decrease. "He finally came up with a solution that was so simple I wondered why I hadn't thought of it myself. His explanation was simply that my results were all wrong."

Despite some teasing about Pincus's tendency to bully his peers and Rock's reluctance to make a splash, the word was out now and both men knew it.

Rock, who always liked to unwind with a drink or two, decided to skip the conference dinner that night and go out with a few other doctors in search of what he called "superior forms of entertainment." When they reached their second drinking establishment of the evening, Rock cut in on a young couple as they danced, leaving the young man to watch as he spun around the room with a much younger woman. When he returned to the gathering of scientists clustered at the bar, Rock presented a complete report on the couple's troubled sex life. He said he had offered suggestions that, if followed, would surely "clear the whole thing up."

Eventually, Rock and his party returned to the lodge at Mont Tremblant. While the other scientists continued to discuss the afternoon's lectures on hormones, Rock and his inebriated crew stripped and plunged nude into the swimming pool.

TWENTY-FIVE

"Papa Pincus's Pink Pills for Planned Parenthood"

"YOU MUST, INDEED, feel a certain pride in your judgment," Margaret Sanger wrote to Katharine McCormick in December 1956. "Gregory Pincus had been working for at least ten years. . . . He had practically no money for his work and . . . then you came along with your fine interest and enthusiasm and with your faith and . . . things began to happen."

Sanger's exuberance sprang from an article in the November 1956 issue of *Science* magazine, the first article on the pill intended for a mainstream audience.

"At last the reports . . . are now out," Sanger wrote, "and the conspiracy of silence is broken."

Their work was far from done, but Sanger and McCormick felt justified in celebrating. Given the ordinary pace of such experimentations, neither of them could have been confident that they would live to see Pincus's work come to a successful conclusion. Now, they were tantalizingly close.

Sanger, however, continued to grow sicker and weaker. She was hooked on sleeping pills and painkillers, and in addition to champagne, she was drinking hard liquor, starting with daiquiris in bed in the morning. At a Population Council conference in New York, she nodded off during the long and repetitious remarks of the Indian ambassador to the United States. When dinner was served, the people sitting beside her tried to wake her but could not. A Planned Parenthood official carried her to her room and put her to bed.

Sanger complained in letters to friends and colleagues that she missed the good old days when the birth-control crusade was a "fighting, forward, no fooling movement, battling for the freedom of the poorest parents and for woman's biological freedom and development." She was still angry that the organization she founded had changed its name from the Birth Control Federation to "that inane one Planned Parenthood." But now, at least, thanks in large part to McCormick and Pincus, she had one more fight, one more "no fooling movement" left to pursue.

Later generations would complain that the birth-control pill put the burden for contraception on women, but that's not the way these women saw it. Sanger and McCormick were born in the nineteenth century. To them, an oral contraceptive wasn't a burden for women. It was a tool. It was an opportunity. And it was on the verge of becoming something more: an achievement that would change women's roles forever.

※

Pincus had the same feeling. There was no precedent for the development of a product like this one, but somehow he was certain he was going to make it work. His nephew, Geoff Dutton, remembered spending time at the Worcester Foundation and in his Uncle Goody's home around 1956. When he was twelve years old, Geoff used to beg Goody to take him to the lab, where he would gaze into the animal cages and dig through the stock room in search of materials he could use to conduct experiments at home.

"Take whatever you want," Uncle Goody would say, and the boy would fill boxes with test tubes, beakers, and bottles of chemicals. "He didn't inspect it," Geoff recalled. "He didn't seem to care what I was taking." One day at about that time, he said, a big box arrived in the mail. It was from Uncle Goody. "In it were bottles of sulfuric acid, nitric acid, a couple bottles of mercury . . . materials you could make explosives with," Geoff said. "And I did make some." While that was Dutton's clearest memory from 1956, he also remembered his parents talking to his uncle about the oral contraceptive Pincus had in development. It was nearly complete, Geoff's parents said, and it was going to be big.

Pincus had enough confidence in Searle's progestin that he began offering it to friends and relatives and getting their informal reports on the drug. Peggy Blake of Morris Plains, New Jersey—whose connection to Pincus is unclear—wrote to him on July 28, 1956, to say she was not happy with the effects of the pill thus far:

> *Dear Dr. Pincus,*
>
> *I have finally started taking SC-4642 (known in this household as Papa Pincus's Pink Pills for Planned Parenthood) and would like to enquire* [sic] *about some possible side effects. The thing is, I have been taking them for 8 days, and during that time have had headaches and some nausea. Also possible slight edema. Do you think it could be the SC-4642? Or have I more likely got some bug or other? Could you let me know your opinion on this soon. I shall continue to take the pills until I hear from you.*

Pincus wrote back saying that her description of the symptoms "pretty much persuades me that the effects you observed are due to the pill." The same symptoms appeared in about 5 percent of all cases, he explained, but they often disappeared or at least diminished as women began their second month of use. "What you should do is, of course, up to you," he wrote. "I don't see much sense in making yourself unhappy."

Peggy Blake did quit taking the pill, and just in time, to hear her tell it, because she had also begun to suffer serious psychological effects. "I was ready to murder anyone who momentarily got in my way, and to burst into tears on practically no provocation." Whether this was directly a result of the pills or simply the anger she felt at suffering so many physical symptoms, she wasn't sure. When she quit the pill after ten days, Blake went to see her doctor to make sure there was nothing wrong with her. "I shall forward you the bill," she informed Pincus, "since, though I am glad to be an experimental subject for you, I don't think I should lose money on the deal."

Blake's letter might have served as a warning, but Pincus was still not terribly worried. Side effects were called side effects for a reason: they were not the main concern. The priority was to make sure no one taking the pill got pregnant. There would be time enough to tinker with the dosages and even the chemical makeup of the pill to see if the side effects could be reduced or eliminated.

※

Dr. Rice-Wray began distributing birth-control pills in early April 1956. She would give each woman a bottle full, enough to last twenty days. "When the bottle is over and you start menstruating," she would tell her patients, "you count one, two, three on your fingers and when you have counted all your fingers that is the time to start again" on a new bottle. Even though she tried to keep the instructions simple, mistakes occurred. At least one patient went home and took all the pills at once. Others shared them with friends. The doctor, nurses, and social workers tried handing out calendars. They tried giving the women beads on a string to help them count. Nothing worked.

Still, Rice-Wray assumed that over time women would get the hang of it.

Soon another problem arose. Less than three weeks into the trials, a reporter for *El Imparcial*, a San Juan newspaper, learned of the experiment and called public health officials for comment. Was it true that a birth-control pill was being tested in Rio Piedras? Rice-Wray received a call from one of her supervisors, asking her to confirm that she was in fact working on a birth-control experiment. Rice-Wray said yes, but she was doing it on her own time, with private funding. The public health office had nothing to do with it. Patients were not being seen in health department offices or by health department personnel. Her boss was skeptical but took her word for it.

The newspaper story that ran the next day began, "A woman dressed as a nurse and who alleges to be working for the state government is distributing . . . some pills to avoid conception and to counteract the increase in population in Puerto Rico." The article went on to claim that Dr. Rice-Wray had "confessed" to directing the project. Her boss was quoted saying he believed it was a "bad combination" for government employees to participate in such work.

After the newspaper story appeared, thirty women dropped out of the trials. Some of them did so because their husbands objected, others because they were worried about what their priests would say, and still others because they were experiencing unpleasant side effects. Soon, a group of Catholic social workers sponsored a program on one of the local TV stations in San Juan in an attempt to discourage women from participating in the trials. After six months, an additional forty-eight women had quit, leaving only about twenty of the original one hundred. The controversy also forced Rice-Wray to resign from her job at the health department, although she would continue working for the Family Planning Association of Puerto Rico. "It's obvious," she wrote in a letter to Pincus, that she caused her boss at the health department "too much discomfort with my planned parenthood activities. I was told by someone . . . 'They respect you but they are afraid of you.'" Rice-Wray worried about how to replace the lost income. She had two children to care for, and she was concerned that she might soon be too old to find a job that would pay well and offer long-term security. She warned Pincus that, while she was committed to their work in Puerto Rico, she would have to drop it if a good job came along.

Over time, the Catholic propaganda campaign backfired and the controversy disappeared, although Rice-Wray would soon announce plans to leave Puerto Rico for a job in Mexico. In the meantime, though, she and the nurse Iris Rodriguez went door to door in Rio Piedras and wrote an article of their own for a local newspaper. In the article and in their conversations with the women of Rio Piedras, they offered reassurance about the safety of the pill while intentionally leaving out mention of the fact that the pill was still experimental. "We will only say that it is being made by the Searle Co. and well known manufacturers, that it is not in the market, that it has proved to be a wonderful contraceptive and that we only have a small amount for special cases," Rodriguez wrote in a May 8 letter to Pincus.

In some cases, women who had not yet heard about the new contraceptive learned about it in church. They would listen on Sunday as their priests made fiery sermons about a forbidden pill, and on Monday they would arrive at Dr. Rice-Wray's office asking what exactly was forbidden and how they could get it.

"Continually they are ringing this office and asking for the pill, going to see Dr. Rice-Wray and calling on me when I make the visits," Rodriguez wrote.

There was another unintended consequence of the Catholic campaign: Many of the women who dropped out of the program quickly became pregnant. As they walked through the streets of Rio Piedras, their stomachs swollen, they served as moving billboards for birth control. Or, as one doctor involved in the testing put it, "their unwanted plight became a major convincing influence for others where they lived."

After a slow start, the study was never again at a loss for volunteers. By the end of 1956, 221 women had participated. Seventeen of those women had gotten pregnant—a fact that might have been troubling to some scientists, but not to Pincus. The pregnancies had nothing to do with the pill, he told Katharine McCormick in a letter following his visit to the island. Women were getting pregnant because they weren't following instructions. They either forgot to take the pill every day or chose not to take it because the side effects were becoming too much to bear. There were two ways to fix that problem. First, he would need to work with the Puerto Rican doctors and social workers to help them better educate the women participating in the experiment and have doctors give them more regular checkups. Second, he would see if he could do something to reduce the pill's side effects.

To Pincus, however, the success stories registered more than the side effects. On his visits to Puerto Rico he met or heard the stories of some of the women taking his pill.

Herminia Alicoa was thirty-two years old, with three children from her first husband and two from her second. Her husband had recently been released from an insane asylum (for the third time) but refused to undergo sterilization. She began taking the birth-control pill seventeen days after delivering her youngest child.

Julia Garcia was thirty years old and had ten children between the ages of sixteen years and ten months. Her husband was sick and he drank heavily, forcing Garcia to perform odd jobs to support her family. Her husband refused to be sterilized or to let her undergo sterilization. Her husband had never allowed her to use any form of contraception and insisted on having sex with her every day. She signed up for trials of the pill because it was the first form of contraception she'd encountered that she could use without her husband's knowledge.

Fanny Quines was thirty years old with five children ranging in age from eight years to sixteen months. Although she was a Seventh-day Adventist and her religion prohibited her from using birth control, she had tried several methods in the past. Since the birth of her youngest child, she had not menstruated. But when she enrolled in the trials and started taking the birth-control pill, her normal periods returned.

Pincus and Rice-Wray were both pleased with the results of women such as these, but Rice-Wray continued to worry about the side effects more than Pincus did. She tallied the numbers: Among the first 221 women in the study, 38, or about 17 percent, reported negative reactions to the drug, and at least 25 women withdrew from the study specifically because of those reactions. There were twenty-nine complaints of dizziness, twenty-six complaints of nausea, eighteen for headaches, seventeen for vomiting, nine for abdominal pain, seven for weakness, and one for diarrhea. Other women complained of bleeding in between their periods, but Rice-Wray said most of those women were able to stop the bleeding by doubling their doses of the pill.

In December 1956, she and Pincus traveled to Skokie, Illinois, to present their findings to top officials at G. D. Searle, who would soon face a decision about whether to offer Pincus's pill to the masses and put their company's reputation and financial future on the line. The company had already patented norethynodrel and had recently trademarked a name for the new drug. They called it Enovid.

Jack Searle, the company's president, sat in on the all-day meeting of scientists. John Rock was there, too, along with Dr. Celso-Ramón Garcia, who was working in Rock's clinic and assisting with the research in Puerto Rico.

Pincus, of course, remained bullish. He was ready to expand the studies with an eye on seeking approval from the Food and Drug Administration. Rock was more cautious. He called the data collected so far "meager" and pointed out that about 20 percent of women participating in the trials reported nausea or breast pain. Rock made no comment, however, on how seriously those side effects should be taken. He was certainly not suggesting that the tests should be suspended.

There was only one woman in the room that day, and that was Rice-Wray. She didn't raise her voice or accuse the men of callousness. She spoke to them in the language they understood best, that of the scientist.

"Enovid gives 100 percent protection against pregnancy in ten-milligram doses taken for twenty days of each month," she said. "However, it causes too many side effects to be acceptable generally." In her view, Enovid was not good enough. Not yet, anyway.

TWENTY-SIX

Jack Searle's Big Bet

"SEX BEFORE MARRIAGE?" Sue Dixon asked with a chuckle. "Well, maybe some of my friends. I was married a virgin."

She gave her husband a warm grin before adding, "But it was nip and tuck."

Sue Dixon's father was Jack Searle, the man who would make the final decision in 1956 as to whether the Skokie, Illinois–based drug company would attempt to become the first in the world to market an oral contraceptive.

Sue's husband, Wes Dixon, smiled across the living room at his wife and elaborated on some of the nipping and tucking that took place in the early 1950s, before they were married. "I took her up to Michigan on a hunting trip and I tried to get into her bedroom," he recalled, "and she said, 'Nope, we don't do that.'" And they didn't. Not until they were married.

Sue and Wes met at a cotillion dance in 1951. They married in May 1953, had their first daughter eleven months later and their first son eighteen months after that. Then, while some of her friends continued to have babies every year or so, Sue took a break. How?

"I went on the pill," she said. "No side effects. Nothing. It was great."

The pill had not yet been approved by the FDA. In fact, Goody Pinçus ("Nice, fun guy," Sue said) was still tinkering with its composition. But Sue and Wes Dixon had no second thoughts about becoming early adopters. "When you're in the pharmaceutical business, your outlook is different," Wes recalled.

Sue remembered discussing the pill with her gynecologist, who was also one of her father's golf partners. Her gynecologist wasn't concerned with her taking an experimental drug, but he was concerned with the social aspects of such an innovation. What would happen, he asked, if men and women could have sex "anytime, anywhere, without having a baby?" What would it do to marriage, to relationships, to the nip and tuck of courting?

It was not a question Pincus or officials at Searle were asking. They were like pioneers pulling wagons west across uncharted lands, focused only on the terrain immediately before them and the promise of something better but as yet only glimpsed far in the distance. Would the pill work? Was it safe? Would the FDA allow them to sell it? Would American women go for such a thing? If the answers to those questions came back in the negative, everything else—including Sue's gynecologist's concerns about sex "anytime, anywhere"—would be moot. It was best not to focus too much on the unknowable.

But Sue Dixon was one of the first women in America to try the birth-control pill for birth control. Her experience not only offered an important clue as to how women would respond to the new contraceptive, it also may have informed her father's decision about whether to put his company's reputation and resources behind one of the most daring and controversial pharmaceutical products in history.

※

In 1956, G. D. Searle was one of the country's smaller drug companies. Gideon Daniel Searle, a Civil War veteran, had founded the business in 1888 with a partner in Omaha, Nebraska, after building a small chain of drug stores in Indiana. Family legend has it that Gideon distinguished himself as a caring pharmacist and a smart businessman by dyeing aspirin in a range of colors and selling them to patients based on their various ailments—pink for headaches, blue for general pain, and so on. Two years after founding the company, they moved it to Chicago. Soon after, Searle split with his partner and started calling his operation G. D. Searle & Co.

Gideon's son Claude, a physician, took over the company in 1909. But it was Claude's son, John G. Searle, who made the business big. Jack, as everyone called him, joined the company in 1923 after studying pharmacology at the University of Michigan. Instead of copying drugs already developed by other companies, Jack Searle hired researchers to design new products. That was his most important contribution early on. In 1934, the company hit on a winner with Metamucil, a laxative, and by 1936 the firm had sales of more than one million dollars.

Jack Searle was a short, slender man with wire-rimmed glasses. Though he looked like a Midwestern conservative, at heart he was a risk taker, and he reinvested his company's profits in state-of-the-art laboratories and manufacturing plants, urging his research directors to be aggressive and come up with more new drugs. No doubt that's why Searle hired men like Pincus, scientific gamblers who lost more often than they won. Jack Searle knew that it took only one big product to change the fortunes of a corporation. He'd seen Smith, Kline & French, one of his competitors, double its sales with the discovery of the new tranquilizer Thorazine. Another competitor, E. R. Squibb & Co., reported that nearly half its sales were coming from products introduced in the past five years, including Nydrazid, an antibiotic used to treat tuberculosis. The drug industry was booming in the 1950s. Americans were spending like never before on their health, and each year brought a new array of so-called wonder drugs. And no drug company in the country was pouring a higher percentage of its revenue back into research than Searle. Jack Searle was determined to hit it big.

Even though it had not scored a huge hit since Metamucil, Searle was getting along nicely, with earnings of $6 million on sales of $26 million in 1955. Its stock was rising, and the family was grooming a new generation of leaders. The next wave included Sue Dixon's husband, Wes, who was appointed an executive vice president in charge of foreign markets in January 1956. It also included Sue's two brothers, Dan (who would go on to become the company's president when Jack Searle retired) and Bill (who would become the director of sales and marketing). But Sue Dixon, despite her college degree and sharp intellect, was not being groomed for a job in the company. In part that was because she was pregnant for a good part of the 1950s, but that wasn't the biggest reason. The biggest reason was that she was a woman, and even women who were unmarried and childless were seldom considered for high-level corporate jobs. That much became obvious to Sue Dixon even when she chose a college and a major. Her brothers were encouraged to study medicine and business; she was pushed into fine arts.

Had she considered trying for a career in science or business, given her family's work? "Me?" she asked. "Women didn't work." Even her fine arts degree proved to have little practical use. She gazed across the living room at her husband and smiled. "I wanted to be a great commercial artist, but that all went down the drain with Wes Dixon!"

Marriage and motherhood ended any chance she might have had for a career, but Sue Dixon nevertheless had an advantage over many of her friends and peers. After giving birth to her first two children, she was able to take time before getting pregnant again, largely because her marriage coincided with the work Pincus was doing for her father's company.

Jack Searle liked the fact that his company was still small enough for the chief executive to be involved at every level. He trusted his research men, Al Raymond and I. C. Winter, but he still liked to sit in on their meetings and ask questions of the chemists and biologists in the room.

Despite their radically different backgrounds (Jack Searle grew up in prosperity and liked to hunt and play golf), he and Pincus got along well. Both men were calm, confident, and bold. At one point in the early days of clinical trials, Searle flew to Puerto Rico with Winter to meet Pincus and Rock and see firsthand how the pill was changing lives. Dr. Garcia said Searle immediately grasped the pill's "sociological implications" and became a champion for the cause. Even before his company began selling the pill, Jack Searle donated money to the cause of population control.

Searle had many reasons to continue pushing the pill toward the market. He had a jump on the competition, which meant the pill, if all went well, might earn a great deal of money. He also believed that this medicine, whatever its flaws, would be a force for good, that it would help women and help the planet.

He was still a business executive, though, which meant balancing risk and reward. On the risk side, there was Dr. Rice-Wray stating publicly that the side effects were too great for this pill to be widely administered to women; there were the scientists who said it would require at least five years of testing to be certain that a daily pill for healthy women would not have long-term, unforeseen dangers; and, of course, there was the Catholic Church, which had already made it clear that it would oppose the use of any such contraceptives.

In the end, Jack Searle decided to go ahead—cautiously.

By 1957, Pincus was preparing an important scientific paper for the *American Journal of Obstetrics and Gynecology* on the results of his human trials in Puerto Rico, which would lend credibility to the cause and take the pill a step closer to reality. Searle encouraged him to proceed. But to be careful, he advised Pincus that it would be best to omit the company's name from the paper.

TWENTY-SEVEN

The Birth of the Pill

AFTER THE MEETING of scientists in Skokie, Pincus and Searle once again tinkered with the makeup of their drug, which was now being referred to among team members by its trademarked name, Enovid (pronounced eh-NAH-vid). Pincus had a simpler name. He called it The Pill, as if it were the only one that mattered. There was no such thing as The Soap or The Vacuum or The Car, but this product was beginning to earn a special identifier in his and others' minds because its tablet form so sharply differentiated it from any contraceptive device that had come before. This wasn't merely an incremental improvement; this was a radical reinvention, and it was exactly what women needed. This was *the pill*, the one they'd been waiting for, the one that changed everything.

By 1957, Pincus and Rock both felt confident that the pill worked and worked safely. Their biggest challenges were to get more women to try it and, at the same time, see if they could do something to reduce the severity of the side effects.

Pincus wasn't as worried as Rice-Wray about how much women were suffering. Some people involved in the Puerto Rican trials attributed the reactions among women in San Juan to the "emotional super-activity of Puerto Rican women," and Pincus continued to believe—or hope—that many of the side effects were merely psychosomatic. To test that theory, he designed a simple experiment. He took one group of women and gave them Enovid with the usual warnings about possible reactions. He took another group and *told* them they were getting Enovid but gave them a placebo instead—along with the same warnings on side effects that he'd given the first group. He also told members of this group that they should use an additional form of birth control to be certain they didn't get pregnant. The third group received the real Enovid and no warnings about side effects.

The first group reported side effects in 23 percent of their cycles. The second group, which received placebos that could not have possibly caused any reactions whatsoever, reported side effects in 17 percent of their cycles. The third group—women taking Enovid with no warnings—experienced side effects in only 6 percent of cycles.

Pincus's experiment violated two basic rules of modern medical research. His patients were not informed of the purpose of the study, nor were they warned of the risks. But the results nonetheless further convinced him he was right—many of the side effects were imaginary, the results of expectation and fear.

Reassured, Pincus set out to recruit more women.

Clarence Gamble, the eugenicist who had been sponsoring research in Puerto Rico for years, offered to fund a second trial. He connected the Pincus team with the medical director at Ryder Memorial Hospital in Humacao, a city thirty-five miles from San Juan on the east coast of Puerto Rico. Gamble had been working with the hospital for more than twenty years. Earlier that same year, he had donated a cautery machine, which was used to perform sterilizations, and donated money to train a doctor in how to use it. He also donated a red Jeep so that hospital workers could navigate the muddy, mottled streets of Humacao as they visited patients.

The job of testing Enovid at Ryder Hospital fell to Dr. Adaline

Pendleton Satterthwaite, the only obstetrician-gynecologist on staff. Like Dr. Rice-Wray, Dr. Satterthwaite was an American-born woman who had come to Puerto Rico as a religious missionary. At Ryder, she delivered about six hundred babies a year. She learned quickly after arriving in Puerto Rico that many pregnant women came to the hospital for two reasons: to have their babies and, immediately afterward, to get sterilized. "O, doctora, opéreme," they would tell her. If not for the incentive of sterilization, Satterthwaite said, most of the women would have delivered at home with midwives. Still, Satterthwaite often refused. Unless the women had three or more children and the consent of their husbands, she would not sterilize. Instead, she would teach them to use a diaphragm.

Now, to test the pill, she chose an area of Humacao called La Vega, which Gamble described as "the most impressive" slum he had ever seen, with no toilets or sewers and housing so crowded there was scarcely room "for a squeezed pedestrian." Gamble hired a woman to take a census of La Vega, going door to door and asking mothers how many children they had, whether they were sterilized, and what type of birth control they were using, if any. Satterthwaite began recruiting in La Vega as well as offering Enovid to the women who had been denied sterilization at the hospital.

She had no trouble finding willing patients, but once again, side effects complicated the work. Women complained of breakthrough bleeding, nausea, and headaches. In examining her patients, Satterthwaite noticed another cause for concern: Women taking the pill for several months appeared to be suffering inflammation of the cervix, the part of the womb whose tip projects into the upper area of the vagina. The inflammation wasn't necessarily harmful, but it might have caused bleeding in some patients. "Whatever you call it," she observed, "the cervix looks angry."

Still, women desperate to avoid pregnancy continued to enroll in the studies. With trials now underway in two Puerto Rican communities, Pincus and his team began work on a third location in Port au Prince, Haiti. The increasing numbers of patients gave Pincus and Searle hope that they might soon have enough information to prove Enovid's safety.

※

Meanwhile, Pincus continued experimenting with different dosages and chemical variations, aiming to reduce the side effects. He tried putting a small amount of antacid in each pill, but it wasn't clear if it made any difference. At one point, Pincus and Chang discovered that Searle's compound had accidentally been contaminated with a tiny amount of synthetic estrogen, also known as mestranol. All along, Pincus had been sticking with progesterone and trying to avoid using estrogen because it was known to cause cancer, and he was worried about the long-term effects of its use. His assumption had been that progesterone was safe for women but that estrogen might not be. When he learned of the accidental contamination, Pincus ordered the drug company to get rid of the estrogen, not only because he thought it might be unsafe but because he thought the estrogen might have been responsible for some of the side effects. But when Searle sent the pure form of Enovid to Puerto Rico, the results were a surprise. Not only did the nausea persist, women began experiencing even more breakthrough bleeding.

Now it dawned on Pincus that the accidental contamination might have been a good thing. With more experimentation, he found that breakthrough bleeding increased when the estrogen dose fell, and nausea and breast pain increased when it rose. He also discovered that when the estrogen levels were too low, the pill was less effective in preventing pregnancy. Now, instead of purifying the pill, he suggested that Searle intentionally make its ten-milligram tablets with 1.5 percent mestranol, which seemed to Pincus like it might be the sweet spot. The side effects did not disappear completely, but they were reduced, and the bleeding all but stopped.

They would stick with this one.

Still, in the first months of 1957, dropout rates remained high and the number of women enrolled in the study low—too low, Pincus believed, to win FDA approval.

Pincus knew he would never be able to test the pill on tens of thousands of women. Even thousands looked like a stretch. But there was no benchmark, no solid number that he knew he would need to get FDA approval. So to make his studies sound more impressive, he stopped talking about the number of women participating in trials. In fact, he stopped talking about women altogether. Instead, he talked about the number of menstrual cycles observed in the experiments. "In the 1,279 cycles during which the regime of treatment was meticulously followed," he wrote in one study, "there was not a single pregnancy."

Pincus wasn't the first to employ this trick. Drug companies often used so-called life-tables to analyze data for drugs that were continually used, arguing that the length of time and the number of doses taken mattered more than the number of people taking them. Given that Pincus couldn't openly test his pill in the United States and continued to have problems finding adequate numbers of women in Puerto Rico, the switch in measurement standards made a difference. Simply put, 1,279 menstrual cycles sounded more impressive than 130 women.

The statistical measures mattered to the scientists and would no doubt matter to the FDA, but to those working in Rio Piedras and Humacao, there were other signs that convinced them the drug was proving effective and popular. Anne Merrill, a laboratory worker for the Worcester Foundation who spent time in Puerto Rico, told an interviewer she was excited by the early results: "I would see these young women with many, many children, although I wouldn't know they were young women until I looked at their records. Some of them looked like what I would think of as great-grandmotherly type, you know? All wizened up and gaunt. . . . I'd look at the records and the woman would be thirty-four, you know, with ten children."

She continued:

To see these women after they'd been on for a year. . . . You know, to get them through one year, without having a baby . . . and here they were hale and hearty! But the exciting thing was that, for instance, some lost their babies in a flood, and they decided they wanted another baby and they just stopped the pills. And then we started seeing the babies that were arriving, and they'd be fine.

Merrill said there was "this terror" among the scientists that only girls or only boys would be born to women who had taken the pill, that the hormones might in some way alter the sex of their babies, or, worse, that the babies might be born with deformities. There was great relief, she said, as the social workers kept track of the babies born and found that they were healthy and that boys and girls appeared to be arriving in roughly equal numbers.

Such anecdotal evidence did not prove the drug's safety, of course, but it gave the doctors and researchers in Puerto Rico increasing confidence that they were doing the right thing.

※

Back in the Chicago suburb of Skokie, Jack Searle and his fellow executives faced a dilemma. The pill was sure to be controversial. It was largely untested and firmly opposed by the Catholic Church. At the same time, it represented a potential goldmine to the company that marketed it first. Here was a medicine women would take more often than aspirin or penicillin, a pill they would take every day, possibly for years, in sickness and in health. If it worked, if it won approval, if it didn't make women sick, and if it became popular or at least socially acceptable, millions of women all over the world might each consume 240 tablets a year at a price of about fifty cents a pill. The numbers were mind-boggling. It would easily be the biggest product Searle had ever brought to market. Best of all, they had paid almost nothing for it up to this point, because Katharine McCormick had carried most of the costs of research and development.

Searle faced difficult questions: What were the rules for testing a pill for healthy people? How far did a company have to go to prove the safety of such a product? Was one year of testing enough to measure long-term effects? Five? Ten?

Pincus urged Jack Searle not to get bogged down. In the end, there was no way to answer such questions because no one had ever done anything like this. There were risks with any drug, but the rewards for this drug in particular were like no other. This was a drug that had a chance to make money, change lives, change the culture, and combat massive world problems such as hunger, poverty, and overcrowding. Pincus made one more important argument: Childbirth was dangerous, too, especially for women who were sick, weak, or starving. There was no way to measure how many lives might be saved by a reliable contraceptive, but even without a measure, that had to be factored into analysis of the drug's risks. Years later, the same argument would be made on behalf of legalizing abortion.

Jack Searle and others within the company had been to the slums of Puerto Rico. They knew that Pincus was right. They knew that it was nearly impossible to quantify the potential benefits of this pill.

Searle's director of clinical research, I. C. Winter—or Icy Winter, as everyone at Searle called him—told the story of a couple he knew that had three "extra children" because the wife hated to get out of her warm bed on cold winter nights to fetch her diaphragm from the bathroom.

"You know," Winter said, "those are the realities."

Pincus—who was a shareholder in G. D. Searle & Co. as well as a contract employee—was so confident that he urged Jack Searle to consider buying one of the companies that manufactured the main ingredients for the pill. If the drug became the blockbuster that Pincus thought it might be, G. D. Searle would sharply boost its profits by taking control of the supply chain.

Within months of Pincus's recommendation, Searle made its first-ever corporate acquisition, taking over Root Chemicals Inc. of Puerto Rico along with its Mexican subsidiary, Productos Esteroides. In a speech to shareholders, Searle said the acquisition would provide an economical source of the raw materials needed to make some of its new hormone products. He didn't say which ones.

Jack Searle had good reason to move fast. Pincus and Searle had decided to use the compound norethynodrel, which had been patented by Searle's own scientist, Frank Colton. Meanwhile, one of Searle's competitors, Parke-Davis, was exploring the possibility of using the other compound Pincus had considered, norethindrone, which had been produced by Djerassi and Syntex. Djerassi urged Parke-Davis to fight, but Parke-Davis and Searle had business ties and didn't want to go to war. Pincus's pill, in Parke-Davis's view, was "small potatoes."

Other competitors might have jumped in. Everyone in the industry knew what Pincus and Searle were up to. The question wasn't whether other companies had the capacity to make a similar drug; the question was whether other companies had the nerve.

"We were a company with an absolutely impeccable reputation," said James S. Irwin, Searle's top marketing executive at the time. But with birth control, he added, "We were going into absolutely unexplored ground in terms of public opinion."

In the end, Jack Searle concluded that the potential rewards outweighed the risks. But he did hedge his bet in one important way. Instead of seeking approval from the FDA for a birth-control pill, the company applied for the approval of Enovid as a treatment for menstrual disorders. "There has been speculation," Jack Searle told the company's stockholders, "that the drug may have use in the field of physiological birth control."

Searle's application to the FDA in 1957 made no mention of contraception. Amenorrhea, dysmenorrhea, and menorrhagia: these were the menstrual problems Enovid was said to combat. The company also claimed that its new drug would be used to treat infertility because, even though the number of cases was small, tests showed that women resting their ovaries for several months were more likely to become pregnant when they stopped taking the drug. It was the "Rock Rebound" again. FDA officials could read between the lines, of course. They could also read in the mainstream press that Pincus and Searle were studying the effects of Enovid as a contraceptive. For the time being, though, it didn't matter. Inspectors couldn't reject the drug as a contraceptive because Searle wasn't asking for approval as a contraceptive. The only question was whether it worked safely and effectively in treating menstrual disorders.

If the FDA approved the pill for menstrual disorders, Enovid would be legal and available by prescription at every doctor's office in the country. That would make it much easier for Pincus to find gynecologists willing to prescribe it for their patients, and if some of those patients were perhaps not exactly suffering from menstrual disorders but in need of an effective form of contraceptive, that would be all right with Pincus. The world would begin to learn what Enovid could do.

TWENTY-EIGHT

"Believed to Have Magical Powers"

ON JUNE 10, 1957, after taking two months to review the application filed by G. D. Searle & Co., the FDA approved the sale of Enovid for infertility and menstrual irregularities. At about the same time, the drug was approved for the same uses in England under the brand name Enavid.

Pincus and the team at Searle were ecstatic, as were Sanger and McCormick. Searle began advertising the drug in medical journals with a picture of a West African fertility doll and this legend: "Believed to have magical powers, these objects are carried by young women of the tribe when they desire to become pregnant." Hoping to avoid controversy, at least for the time being, Searle advertised the drug as a boon to infertile women. Enovid, they said, would help women regulate their cycles. Once they achieved better control of those cycles, the company promised, they would have a greater chance of becoming pregnant.

That was the official line. Let women discover Enovid; let them see that it was safe. Let doctors prescribe it. Let everyone discover that it worked—as a regulator of menstruation and, oh, by the way, as a contraceptive, too. As the word spread from woman to woman, doctor to doctor, Searle would quietly make the necessary preparations to market the drug as the world's first scientific birth-control pill.

Pincus couldn't wait. In June, he went to Sweden and boasted in a widely reported speech that he had developed "an almost 100 percent effective pill for preventing pregnancy." Rock, still uneasy about incurring the wrath of the Catholic Church, hurriedly sent him a telegram. "SUGGEST BUTTONING UP," it read.

Pincus did not button up. Instead, he gave an interview to the science writer Albert Q. Maisel for an article that appeared in the August issue of *Ladies' Home Journal*, which had a circulation of more than four million, making it the most popular women's magazine in the world and one of the most popular magazines of any kind. Maisel offered readers a detailed account of the "brilliant and painstaking work" Pincus and his team were doing to develop what the three-page article referred to as "the new hormone preparation . . . for widespread and long-term use as a method of controlling fertility."

Officials at G. D. Searle were furious. Six months earlier, Searle had sponsored a symposium for some of the nation's leading reproductive scientists and gynecologists, hoping to educate them about Enovid and win their support for the drug. The company had promised that the material presented at the symposium would not be released to the public. Now, scientists who had attended the meeting wrote to Searle, asking how details from the meeting had made it into print—and into *Ladies' Home Journal*, of all places. I. C. Winter wrote to the scientists in apology, saying the Searle company had neither instigated the article nor cooperated with the author. But Winter knew who had. Goody Pincus and the writer Maisel were old friends. In fact, not long after the article appeared in *Ladies' Home Journal*, Pincus would put Maisel on the Worcester Foundation payroll to help with publicity.

※

Pincus had done it again. Just as he had twenty years earlier at Harvard, when a *Collier's* article had helped sink his academic career, he had used the mainstream press instead of a scientific journal to inform the world about his thrilling, frightening, utterly unfinished work. This time, though, the stakes were even higher. G. D. Searle and Katharine McCormick had spent hundreds of thousands of dollars backing the birth-control pill, and Searle had recently received FDA approval to sell it. It didn't matter that the approval was not for a birth-control pill but for a pill to regulate menstrual disorders; what mattered was that Enovid did the work of a birth-control pill and it was now legal. Searle officials were deeply worried that there might be a backlash against the company, and that such a backlash could lead to a boycott of Searle products and a drop in the company's stock price.

Searle proceeded cautiously. Beginning in July 1957, the company offered Enovid only to doctors on the West Coast. The company advertised the pill for the regulation of menstruation, but salesmen made no secret of its other benefit. Federal law regulated the behavior of drug companies, but not that of doctors. Once a drug was approved for any purpose, physicians were free to prescribe it liberally (medical professionals refer to those other purposes as the "off-label uses").

"Presumably," Pincus told Sanger in a letter, "any physician may prescribe it for any purpose which he considers valid."

In other words, with the arrival of Enovid, a pill for birth control was on the market and available by prescription for the first time, even if it was not yet official and not yet openly referred to as a birth-control pill.

That, however, did not mean Pincus's work was done.

"While we are convinced of its contraceptive action," Pincus said, "Dr. Rock and I are not ready to recommend its universal use." They were still concerned with long-term health risks and short-term side effects. Studies on the long-term effects would take at least another year or two, he said. But there was good news: If doctors all over the country began prescribing Enovid, whether for menstrual disorders or for birth control, Pincus and Rock would soon receive more feedback on the drug's effects than they could ever hope to glean from ongoing studies in Puerto Rico. They wouldn't get blood and urine tests from these new customers, but they would receive anecdotal evidence. If the pill worked, women would tell their friends, and sales would take off. If it did not, or if the side effects were too great for many women to tolerate, Searle would see a drop in orders and Pincus and Rock would no doubt hear directly from many of the prescribing physicians.

Convinced he was on the brink of a great accomplishment and that his progestin-based pills really might be the world's answer to the problem of birth control, Pincus began spreading the word. He used the connections he had long been building as the leader of the Laurentian Hormone Conference, and he encouraged scientists around the world to launch their own experiments with Enovid. Around this time, he made a trip to Paris to give a lecture on his latest findings. When the sponsor of his lecture presented him with an honorarium, rather than taking the check home and depositing it in the bank or turning it over to the Worcester Foundation, he decided to give himself a treat. He applied the money toward the purchase of a silver Citröen DS 19, acclaimed at the time as the most technologically advanced car in the world, with power steering, power disc brakes, and a hydropneumatic suspension that would raise the car ten inches off the ground with the tug of a lever. The Citröen was sleek, sexy, and looked like something out of the future. Pincus had it shipped home, where he became even more of a terror on the roads of Worcester and Shrewsbury.

Pincus's days of hardship were long behind him now, and it was apt that he was driving faster than ever in a car that called attention to its driver. His work on the birth-control pill was a full-throttled race to the finish. But in the 1950s, such bold and hurried approaches to new medicine were not unusual. To receive government approval for a new drug, the manufacturer needed to submit the results of animal and human testing, as well as materials showing how the drug would be marketed. But the FDA, overwhelmed at the time by requests for new drug approvals and badly understaffed, paid more attention to some things than others. It looked carefully to see if drugs were effective, for example, but less carefully at whether the drugs were safe. In part, that explains why all along Pincus wanted to keep doses of the pill at ten milligrams, which he knew was perhaps higher than necessary. Above all else he wanted to be certain Enovid would prevent pregnancy. The FDA's final decision, he understood, would not be based on how many women were tested or on how many of those women experienced side effects. The key was the pill's success rate, and Pincus was determined that the rate should be very near 100 percent.

If the pill was more effective by a wide margin than any other form of contraception, Pincus felt confident the FDA would approve it. The agency approved other drugs with known side effects, including penicillin, which caused life-threatening allergic reactions in some patients; tetracycline, which interfered with bone growth and discolored teeth; and Dilantin, which was known to cause heart irregularities. In all those cases, the rewards were deemed greater than the risks. Pincus believed the same standard would apply to Enovid.

To help encourage doctors to prescribe the pill, Searle sent a letter in July to hundreds of obstetricians, gynecologists, and general practitioners, informing them that Enovid was on the market and that it might do more than its advertisements suggested. "There is adequate evidence," the letter said, "to indicate that the drug will inhibit ovulation when the physician so chooses and that it is safe for this purpose in short term medication."

It was not the ideal way to get feedback. There was an obvious risk involved. If the drug did serious damage it would be too late for many of the women using it.

※

On October 1, 1957, a few months after Enovid went on sale in the United States, another so-called wonder drug reached the market in

Europe. It was advertised as a sleeping pill so safe and free of side effects that even pregnant women could safely use it. In fact, it also had the benefit of reducing morning sickness, so pregnant women began using the drug even if they weren't suffering from insomnia. In some European countries, the medicine became more popular than aspirin.

Within months, a few physicians began to report problems. Some elderly patients using the new drug experienced giddiness and a slight disturbance of balance. Others suffered dizziness, trembling, and cold hands and feet. But the drug's manufacturer, Chemie Grünenthal, shrugged off the reports, assuming that there was nothing seriously wrong with their hottest-selling new product, that it was doing far more good than harm, and that in the end it would prove perfectly safe.

But thalidomide proved far from safe.

In the years to come, thousands of children around the world whose mothers had taken the drug while pregnant were born with serious physical disabilities, including flipper-like arms and legs. Many of them were abandoned by their parents and institutionalized. Others had their flippers amputated so they could be fitted with prosthetic arms and legs.

But as Enovid went on sale in the United States, the horrors of thalidomide had yet to be discovered. The age of the wonder drug rolled on, unblemished, as Enovid got its chance.

※

The article about birth control in *Ladies' Home Journal* triggered a chain reaction. Soon there were articles in *Science Digest*, the *Saturday Evening Post*, and newspapers all over the world. The waves of publicity did more than rouse the ire of officials at Searle and leaders of the Catholic Church. It sent women in search of the drug, which Searle was still in the process of rolling out. Some of them, frantic, appealed directly to Pincus for help.

A woman from Indianapolis wrote:

I'm about 30 years old [and] have six children. . . . We have tried to be careful and tried this and that, but I get pregnant anyway. When I read this article I couldn't help but cry, for I thought this is my ray of hope. As of yet can these anti-pregnancy pills be purchased? Where—How can I get them? Please help me. . . . I beg you please help me if you can.

A man who worked at the University of Chicago wrote, "To save my married daughter from risking her life by making 3 times of artificial abortions during last year's time, I beg you to send her some of the tablets."

"I absolutely need your help," wrote a thirty-year-old woman in Canada who was pregnant at the time with her fifth child. "I do not think that I am fitted to raise 10 or more children—it costs too much and my husband is not giving me much help to educate my [here she crossed out "our"] children. . . . Please help me!"

But if officials at Searle were angry with Pincus for attracting too much attention, their anger didn't last. It quickly became clear that demand for the new drug was greater than anyone had imagined.

For young women, the arrival of Enovid in 1957 was a major event. All their lives women had been told birth control was illicit, forbidden, a violation of nature. Now, the federal government declared that if you had an irregular cycle—and every woman could reasonably claim her cycle was irregular—you could get a prescription for a pill that would make you regular and also protect you from pregnancy.

Some doctors asked more questions than others, and some asked no questions at all. But women—married and not—quickly learned which physicians were willing to write prescriptions and made a beeline to their offices.

Men noticed, too.

"I remember . . . when it became legal," Hugh Hefner said. "It made a big impression on me at the time. I recognized the importance of it. . . . I recognized it as exactly what it was: a powerful weapon."

In the years ahead, Hefner would argue in the pages of *Playboy* that this weapon should be used to launch a sexual revolution, making intercourse an expression of love instead of merely a means of procreation. Asked if he could remember the first time he had sex with a woman who was taking Enovid, Hefner laughed. "I don't remember that," he said. "The nature of the pill is such that a guy would not necessarily be focused on that. Of course, from 1960 on I think all the women I knew were on the pill."

The weapon was being put to use with verve, with joy, with excitement, trepidation, passion. Searle did not have to market the pill as birth control because men and women were learning for themselves what it could do. It didn't hurt, either, that the FDA had required the drug company to include a warning on each bottle that said Enovid prevented ovulation. In other words, the real purpose of the drug was listed as if it were a side effect.

As it turned out, said Searle's I. C. Winter, "It was like a free ad."

TWENTY-NINE

The Double Effect

"GOOD EVENING," THE TV newsman said, his shoulders squared, eyes locked on the camera, cigarette smoke drifting toward his resolute face. "What you're about to witness is an unrehearsed, uncensored interview on the issue of birth control. It will be a free discussion of an adult topic. . . . My name is Mike Wallace, the cigarette is Philip Morris."

It was September 21, 1957, an unusually warm and humid day in New York City, when Sanger arrived at the studio to be interviewed. Wallace, her interviewer, was thirty-nine years old, a former radio and stage actor, and had risen to fame on television a year earlier as a confrontational interviewer who seemed to enjoy making his interview subjects, and perhaps even his audience, uncomfortable. He described his own style as "nosy, irreverent, often confrontational." Wallace might have liked Sanger. They both were tireless combatants, after all, as well as massive egotists. But Wallace wasn't looking for a friend as he sat down with her for their interview; he was looking for a fight. Sanger's chief appeal to him at the time, he said, "was as a figure of controversy, an intrepid gadfly who had the temerity to challenge the moral authority of the Roman Catholic Church." Wallace came out swinging, and Sanger, appearing uncomfortable on camera, stepped into every punch.

Wallace asked long and complicated questions, at times quoting newspaper articles and Sanger's personal letters that dated back decades, and asking her to respond to controversial comments she or others had made. Sanger pursed her lips and nervously scratched at her neck and scalp. She had difficulty completing her thoughts at times.

Wallace began by accusing Sanger of abandoning her children and husband as a young woman because she craved "this joy, this freedom" that the birth-control movement offered. After that, he began pushing Sanger to explain her religious beliefs. Did she believe in God? In sin? Did she consider infidelity a sin? Divorce? Murder?

Sanger tried to respond assertively, saying she believed the greatest sin was bringing unwanted children into the world. But Wallace wasn't having it.

"May I ask you this?" he continued. "Could it be that women in the United States have become too independent—that they followed the lead of women like Margaret Sanger by neglecting family life for a career? Let me quote from your biography describing your second marriage to Noah Slee. Quote, 'In New York Mrs. Sanger maintained every clause of their compact of independence. They had separate apartments—they telephoned each other for dinner or theatre engagements or passed notes back and forth.' Would you call this a sound formula for marriage Mrs. Sanger?"

"For different people, yes," Sanger said, "it certainly was for me, and for my husband. We had a very happy marriage. . . . He had different friends than I had."

Television was new, and Americans, including Sanger, were only beginning to understand its power. At the start of the decade, televisions were in 9 percent of homes. By the end of the decade they were in 90 percent, and the average person was watching forty-two hours a week. Back in Tucson, Sanger's teenage granddaughters watched in dismay, waiting for Sanger to find her footing and launch a counter-attack. They were not used to seeing her bullied, seeing her seem so uncertain of herself. Her son Bill, watching in New York, began to cry.

Wallace may have won the battle, but he had missed the story. He never asked Sanger about the birth-control pill. If he had, he might not only have broken news, he also might have engaged Sanger in an even more heated discussion about whether sex should be for pleasure and whether such a pill might encourage greater numbers of women than ever to "neglect their families" in favor of careers. But Wallace did capture the moral complexities of Sanger's life and work. Though she had stated repeatedly for more than forty years that she was interested in using science and politics to solve social issues, her crusade had always been a moral one, and her whole life had been an expression of that crusade. In her war with the Catholic Church, in her fight for greater sexual freedom, and in her belief that women would never be treated equally until they controlled their own bodies, she had never shied away from battle and stayed in for the long war.

Science was forging ahead, largely without Sanger now, in the search for better birth control. A new generation was thinking in fresh ways about what women needed to do to gain power and claim independence. Sanger was a part of history. She had fought and compromised, making bold choices and bad ones, but she was no longer the leader of this movement or any other.

As the harsh lights of the television studio cast shadows in close-up across Sanger's wrinkled face, American television viewers watched a warrior whose time had almost passed.

※

After Sanger's appearance, sacks of mail arrived at ABC's offices and at Planned Parenthood. Sanger glanced at a few—"It is just too bad that your mother didn't practice birth control 95 years ago when she had you—(because you look that old you old bag)," one read—and threw the rest away. But she did read an editorial that appeared in response to her TV interview in a September issue of the *The Evangelist*, a weekly publication of the Catholic diocese in Albany, New York. It read, in part: "Wallace, who claimed 'to explore the economic, moral and religious aspects of birth control' was the instrument whereby Mrs. Sanger, veteran proponent of barnyard ethics and race suicide, was given her entrance into millions of decent homes to taint them with her evil philosophy of lust and animalistic mating."

Sanger wrote in her diary that the "R.C. Church is getting more defiant and arrogant. . . . Young Kennedy from Boston is on the Stage for President in 1960. God help America if his father's millions can push him into the White House."

As usual, Sanger had nothing good to say about Catholics and the Church. She had accepted John Rock, but that was as far as she would go. Had she been less obstinate, she might have realized that John Fitzgerald Kennedy, the Democratic senator from Massachusetts, was part of a new generation of Catholics. Kennedy's beliefs had been shaped by his religion, but religion had not had a strong effect on his political views. Had Sanger been less obstinate, she also might have noticed that by the late 1950s some theologians were openly expressing curiosity about the birth-control pill. As the scientific experiments surrounding the pill garnered publicity, Catholic women were meeting with their priests and asking whether this new form of birth control might be acceptable to the Church. Responses varied. Priests all over the country had to make up their own minds as to what to tell their parishioners. Not all of them stuck to the Vatican line.

When Enovid went on the market in 1957 and doctors began offering it clandestinely to women for contraception, everything changed, in part because the Church had not yet figured out how to react. Here was something new. It wasn't billed as birth control and yet it worked as birth control. Even better, it didn't *look* like birth control. A woman who was ashamed or uncertain of her anatomy did not have to struggle with awkward devices to have sex. She could take the pill in the morning and forget about it entirely when she decided whether or not to make love to her partner that night. She could take it without telling anyone but her doctor. It was the most personal of choices, and that important change in psychology allowed many women to consider contraception for the first time. Often their friends encouraged them, their doctors helped them, and their priests, receiving no clear guidance from the Vatican, did nothing to get in the way.

John Rock was not the only one who thought the Church might eventually approve the use of Enovid. The Church allowed women to have sex after they'd undergone hysterectomies. That was because their sterility had come about as a cure for a disease. If a woman had undergone a hysterectomy for the express purpose of having sex without fear of making babies, the Church would not have condoned it. In other words, the Church had declared that it was acceptable for women to shut down the process of ovulation to treat endometriosis, excessive bleeding, or painful menstruation. Enovid treated those same ailments in much the same way, by shutting down ovulation.

Pope Pius XII appeared to accept this line of thinking, at least to an extent: "If a woman takes this medicine, not to prevent conception, but only on the advice of a doctor as a necessary remedy because of a disease of the uterus or the organism," he said, "she provokes an indirect sterilization, which is permitted according to the general principle of actions with a double effect."

The principle of double effect, introduced by Thomas Aquinas, is often invoked to explain why actions that can cause harm may be permitted. It is permissible to kill a man in self-defense, for example, provided one does not intend to kill him. The action is acceptable because the goal was saving one's own life, not causing harm to one's assailant. A doctor who believed that abortion was wrong, even to save the mother's life, might be willing to perform a hysterectomy on a woman with treatable cancer because the bad effect (the death of her fetus) flowed as an indirect result of the good effect (saving the mother's life). Intent was everything. Evil was still evil. But as long as the evil was unintentional, guilt might be allayed.

Not surprisingly, the debate grew murky. Was an irregular menstrual cycle enough reason for a woman to take a pill and receive both the good effect (regularity) and the bad (contraception)? Was painful menstruation painful enough to qualify as an ailment? Was fear of pregnancy a legitimate medical concern? What if the fear of pregnancy was causing a woman to suffer a mental breakdown or raised her risk of a heart attack? What if pregnancy might cause her to suffer heart failure? What if pregnancy might cause her other children to starve?

Then there was the matter of the rhythm method. If the Church permitted women to use the rhythm method to determine when it was safe to have sex without fear of pregnancy, perhaps it would permit women to use the pill to regulate their menstrual cycles and practice safe sex more reliably.

"It seems to me," the American Jesuit John Connery wrote, "that perfect regularity is as legitimate a goal as perfect health or perfect vision."

These were not easy problems to solve, especially for a group of celibate men. But in the end, only one man's opinion would matter when it came to setting the Church's policy, and that was the pope.

THIRTY

La Señora de las Pastillas

PINCUS AND HIS wife Lizzie never explained sex to their daughter, Laura. Instead, when Laura was about fourteen, her father left a book called *A Marriage Manual*, by Dr. Abraham Stone, on the dining room table for her to read.

"I thought it was so strange," she recalled.

In the fall of 1957, when she was twenty-two, Laura went to work for her father in Puerto Rico, signing on as an administrator for the continuing field trials in Rio Piedras and Humacao. Laura was a great beauty, like her mother, with an hourglass figure and penetrating eyes. She also had her mother's pluck. She was still a virgin at the time, but before her work in Puerto Rico was done she too would begin taking the pill her father had invented.

"La señora de las pastillas," the women in the Puerto Rican trials called her. The pill woman.

In one of the slums Laura visited, urban planners were building new houses over the existing shacks so that residents could continue to live in them during construction. She was fascinated by the maturity of the young women she met, many of whom were younger than she but already caring for children.

She recalled:

Many of them were never actually married. They were too poor to get licenses. They would pose with their heads through cardboard wedding dresses for their wedding pictures. I would ask them, as Catholics, how did they feel about not being married, and they would say, "I have a direct relationship with God, and he understands why I have to do this." They were very poor. They were desperate not to have more children. But the men, some of them were macho, and they wanted as many children as possible. With the pill, they didn't have to tell their husbands they were using birth control.

For Laura, who had recently graduated from Radcliffe College, Puerto Rico was another kind of education. She marched into the slums with her clipboard in hand, asking questions such as "How many times did you have intercourse this week?" and "Did you practice interruption as a method of birth control?"

Laura spent a year in San Juan, and while she was there she met Michael Bernard, a young Harvard graduate working as an urban planner for the Puerto Rican government. When Laura told her father she had fallen in love and begun taking the birth-control pill, he was not upset. His main concern was side effects.

She said she hadn't experienced any.

※

Throughout the late 1950s, Goody continued to make frequent trips to San Juan, with occasional visits to Haiti as well. Lizzie usually accompanied him. They would stay at lush tropical hotels and sit on the patio at night, smoking, sipping cocktails with little umbrellas in them, and listening to the surf crash against the shore. In the mornings, Lizzie would sleep while Goody would meet with the doctors, scientists, and nurses working on the trials.

Pincus usually brought laboratory technicians from Massachusetts to Puerto Rico so that they could perform Pap smears and endometrial biopsies, tests that the medical team in Puerto Rico couldn't handle.

The trials in Rio Piedras and Humacao were proceeding smoothly, but efforts in Haiti were not so successful. The team of doctors and social workers there was not as competent, and high illiteracy rates made it more difficult for Haitian women to follow directions. Still, Pincus was determined to enroll as many women as possible in the study, and to keep a close eye on those who'd been participating the longest. He was also eager to test lower doses of Enovid—2.5 and 5 milligrams instead of 10—to see if it would remain effective while reducing side effects.

By the end of 1958, more than 800 women had enrolled in tests of the pill, but only about 130 had taken it for a year or more. Those were small numbers for a medical trial, even by 1950s standards. Yet while some doctors and journalists had suggested it would take years of testing on thousands of women before a drug like this would be approved for contraception, officials at Searle did not want to wait for Pincus to obtain more results. In December, even as Pincus continued to tinker with the dosages and the makeup of the pill, Searle was preparing to make the next big step: requesting approval from the FDA to sell Enovid as an oral contraceptive.

The initial approval had been quick and easy, but only because the company had avoided calling the drug a birth-control pill. Now, Searle was hoping to avoid controversy and gain another speedy victory by filing a supplemental application, essentially making the case that a new use for the drug had been discovered and asking for permission to advertise that use rather than starting over and applying as if this were a new drug. It was a standard procedure in the drug business, but there was nothing routine where Enovid was concerned. This was not a cure for a disease. It was not a treatment for a condition or pain. This was a drug designed to change the way women lived. What's more, it was a drug that women might use for twenty years or more. No one had the slightest clue what kind of long-term side effects it might produce or what standards should apply in determining its safety.

Pincus continued to insist the pill was harmless, and Searle's scientists agreed. Meanwhile, Jack Searle crunched the numbers. Contraception was an industry worth $200 million in the United States, with condoms accounting for $150 million of that amount and diaphragms and jellies representing the next largest category at about $20 million. That was $200 million going to inferior products. The pill had the potential to be bigger than anything currently on the market. But the big profits would come only if Searle got to the prize first. In 1958, *Fortune* reported that the Ortho Pharmaceutical Corporation, the industry leader in diaphragms and jellies, was shifting the majority of its research budget to the search for oral contraceptives, trying to catch up with Searle.

"It is no news, I am sure," I. C. Winter wrote to Pincus on December 29, 1958, "that the powers that be are breathing down our neck in the hopes of speeding up our application."

THIRTY-ONE

An Unlikely Pitch Man

IN 1958, SEVENTEEN states still had laws banning the sale, distribution, or advertisement of contraception. In Connecticut, it was a crime to "use any drug, medicinal article, or instrument for the purpose of preventing conception." In Massachusetts, where so much of the pill work was being done, it was still a felony to "exhibit, sell, prescribe, provide, or give out information" about contraceptives.

But gradually, one state at a time, the laws were being overturned. Those in place were largely unenforced. It was clearer than ever that a majority of Americans favored some form of birth control, and there were no modern-day Comstocks raising hell about it. On the contrary, there were calls for the government not only to accept contraception but to step in and regulate it. Too many bootleggers were still selling worthless products, and too many quacks were still performing unsafe abortions.

Now it wasn't only crusading sex advocates like Margaret Sanger angling for a birth-control pill; giant corporations and a growing number of politicians and religious leaders were pushing for its acceptance.

For Searle, that meant a market had emerged. Moral and legal issues mattered, but making money mattered, too. The company stepped up to meet the growing demand, confident that the law would find a way to keep pace with the rapid cultural changes taking place all over the country. In a sense, Searle may have been fortunate that Margaret Sanger was old and sick and perhaps somewhat shell-shocked from her encounter with Mike Wallace. Hers was not the face the drug company wanted to present as it prepared to market its newest product. Neither was Pincus's, for that matter.

John Rock, on the other hand, was perfect. He was Catholic; he was tall, silver-haired, and handsome; and he was no one's idea of a flamethrower. He was married with children and grandchildren, and he possessed one of the finest reputations in all of American medicine. Searle, a family-owned, Midwestern company, could not have hired a better spokesman if they had issued a Hollywood casting call and auditioned thousands. As the race for the birth-control pill came close to its finish, the company put him to use.

Rock was receiving about ten thousand dollars a year (eighty thousand by today's standard) in grants from Searle, but it was not simply money that motivated him. He believed in Enovid. He believed in the birth-control pill. He believed it was safe. He believed it would be a great boon for women and a help to marriage. He believed the Catholic Church ought to embrace it, not only for the sake of women but also for the sake of the Church. For all of these reasons and others, he happily cooperated when Searle asked him in 1959 to help decide how the company ought to market the pill as a birth-control device—assuming, of course, that the FDA approved it.

In 1959, drugs came with labels and nothing more. There were no pamphlets in the packaging to tell patients how the drugs should be used or to warn them of possible side effects. If patients needed instructions beyond those printed on the bottle, they asked their doctors. For good reason, Searle was more worried than usual about how this drug would be received. For one thing, women would be taking it by choice rather than necessity. They would be trying it as a replacement for other forms of birth control, not to ease a pain or cure an illness. Searle officials recognized that it was important for doctors and patients to feel comfortable and informed as they tried the pill for the first time. They wanted to set the right tone, striking a balance between the pill's medical uses and its social benefits, and they wanted to make sure the message about the drug's effects was delivered in the plainest and least sensational way possible. That's where Rock came in.

"I'm not experienced in this kind of promotional work," Rock wrote to an executive at Searle, adding, "I don't know if you will approve" of the results. But he agreed to try. He suggested the company use phrases such as "child spacing," "postponement of pregnancy," and "suppression of ovulation" in its literature rather than "contraception" or "birth control." Searle would settle eventually on "family planning" as its euphemism of choice.

In a brochure prepared for doctors, Rock spent a page and a half describing the menstrual cycle for general practitioners who were perhaps not as familiar with it as gynecologists were. Enovid, he wrote, "completely mimics" the action of natural progesterone in suppressing ovulation. In a separate brochure for patients, he wrote even more plainly: "Enovid is an artificially made hormone that is chemically quite similar to the two hormones, *estrogen* and *progesterone*, naturally produced in the human ovary." Although the pill was ten times more powerful than progesterone, he wrote, its action was "quite like that of the natural hormone." He never used the words "contraception" or "birth control," but he did repeatedly use the word "natural," in part because he was writing not only for women but for the Catholic Church officials who might be reading the same materials. Once women embraced the pill, as Rock believed they would, he hoped the Church might follow.

In Rock's mind, the pill served as an extension of nature, and he was not the only one who thought so. James Balog, a researcher at Merck and Company, said that Searle and Rock were on to an important idea.

"I felt that the pill might be the theological way out," Balog told the writer Bernard Asbell. "It's not abortive because there's no ovum, you're not interfering with the will of God by putting a spermicidal jelly on that poor little sperm who's trying to find this wonderful little ovum to do his thing. . . . If you believe the soul is created when the sperm unites with the ovum, there is no fertilized ovum, therefore no soul."

There was a real sense that the pill might so thoroughly reinvent the notion of birth control that even the Church would find it impossible not to go along with it. In that regard, John Rock never saw himself as a radical; he believed his objective was a realistic one. He wasn't trying to attack the Church. He was merely hoping to lead it in the direction that he so strongly believed would prove to be the right way.

※

In 1959, transatlantic jet service became widely available to commercial passengers. Americans also greeted the first astronauts, a group of fit and handsome U.S. test pilots known as the Mercury Seven. At the age of forty-two, John F. Kennedy emerged as the undeclared Democratic frontrunner in the next presidential election. Barney Rosset, owner of Grove Press, sued the government to overturn censorship and obscenity laws in order to publish D. H. Lawrence's *Lady Chatterley's Lover.* Philip Roth published *Goodbye, Columbus*, a novel in which the plot turned on a young woman's decision to visit the Margaret Sanger Clinic and get a diaphragm—and her mother's subsequent discovery of the device. As the pain of World War II and the Korean War faded and a wave of prosperity swept across the country, Americans began to push for all kinds of change, pressing the limits of what had previously been possible socially, economically, and morally.

In *Thy Neighbor's Wife*, Gay Talese reported that, in 1959, "after a Chicago vice squad had arrested 55 independent news vendors for selling girlie magazines, a jury of five women and seven men—uninfluenced by a church group that sat in the courtroom holding rosary beads and silently praying—voted to acquit the defendants. After the verdict had been announced, the judge seemed stunned, then slumped forward from the bench and had to be rushed to a hospital. He had had a heart attack."

The times were changing. Katharine McCormick found out firsthand when, in the summer of 1959, she stepped into a drugstore in Santa Barbara, California, and handed the pharmacist a prescription for Enovid. She was merely picking up the pills for a friend, but it didn't matter. She had waited many years and spent great sums of money in the hope that such a simple transaction might one day be possible, and that day had come.

Yet for all the dramatic cultural and political change, G. D. Searle & Co. still wasn't certain that women would feel comfortable talking to their friends about birth-control options and asking their doctors for Enovid. In an attempt to find out, the company's top public relations man, James W. Irwin, phoned a few magazine writers and editors he knew, encouraging them to produce stories about his company's new pill and the big changes for family planning it might deliver.

He warned the editors that they might catch heat for the stories. The Catholic Church might protest their magazines. Subscribers might cancel. Then, as the articles appeared, he waited for a backlash that never came.

Life magazine ran pictures of the smiling Pincus and Rock along with a lengthy article that described the pluses and minuses of their newly invented pill. On the plus side, reported the magazine: "Treatment consists of nothing but swallowing a pill once a day, beginning on the fifth day after menstruation and continuing for 20 days. . . . If ovulation is wanted, the patient has only to stop the pills for a while to revert to normal." On the minus side: "Although this seems to be simplicity itself, numbers of women in the clinical trials found it beyond their mental and emotional capacity, and it must be presumed that numbers of Indians, Chinese, *et al.* would likewise."

The *Time* article and others like it made little mention of side effects and even less mention of potential long-term hazards. The magazine editors treated the story of the birth-control pill as they might have treated the story of residential air conditioning or any other great life-changing invention finding its way into American homes in the 1950s. It was a subject of fascination, a glimpse of the future, and another glittering example of the wonders of American ingenuity. Pincus, Rock, and officials at Searle were hit with another wave of letters from women who wanted the pill and wanted it now. The so-called Baby Boom was underway, and many women had already seen enough of it. The more savvy among them knew that help was already available. All they needed to do was find a doctor willing to prescribe Enovid for menstrual regulation.

Though it was not yet officially recognized as birth control, more than five hundred thousand women were taking the pill.

※

On July 23, 1959—two days after Grove Press won permission to publish *Lady Chatterley's Lover*—G. D. Searle made it official and asked the FDA to approve Enovid for birth control. At the time, no one could have recognized that these two small ripples presaged a tidal wave that would sweep away the nation's culture of restraint.

※

Though the number of women enrolled in clinical trials for the pill remained relatively small, Searle submitted the biggest new-drug application in American history at that time, with twenty volumes of data. The application included details from Pap smears, biopsies, temperature readings, and much more. Results for each woman in the study were provided, as well as summaries of all the data presented as both raw statistics and in a dizzying array of charts. This was the moment Pincus, Rock, and others like Rice-Wray and Chang had been working toward, and they put everything they had on the table.

In 1959, the United States possessed one of the most stringent new-drug approval processes in the world. Drugs were originally sold and prepared by apothecaries and pharmacists, and regulation was nearly impossible. By the 1800s, some countries, including the United States, established pharmacopoeias to impose quality standards for medications. These national certification boards did not pay attention to safety or efficacy, but they did address issues of honesty in advertising and purity of ingredients. When con artists began peddling pills that promised to cure cancer, syphilis, diabetes, and more, governments began to enforce more stringent rules. In the United States, the first set of meaningful laws was passed in 1902 after twelve children in St. Louis died from a diphtheria vaccine that had been contaminated with tetanus. The St. Louis health department had made the vaccine. In response, a drug laboratory was created within the Department of Agriculture. That lab would eventually become the Food and Drug Administration.

Still, the U.S. government had little power to protect customers from bad drugs until 1938, when Congress passed the Food, Drug, and Cosmetic Act. For the first time, new drugs had to be approved for safety by the FDA and labeled with directions for safe use.

More than 6,000 new drug applications flooded the FDA in the decade that followed, and another 4,200 came in between 1950 and 1959. The FDA approved more than two-thirds of all applications over the course of the decade. But in all those years and all those applications, the agency had never approved anything like Enovid.

THIRTY-TWO

"A Whole New Bag of Beans"

PINCUS AND ROCK still weren't sure their pill would be safe for women to take for years on end. In fact, even as the FDA began reviewing the field trial data, the scientist and the physician continued experimenting with lower doses to see if the pill would remain effective. In an acknowledgment that their team did not yet have all the data they would have liked concerning long-term health effects, Searle asked the FDA to approve Enovid for no more than two years of use per patient.

In 1959, the FDA had only four full-time physicians and four part-timers assigned to investigate the 369 new drug applications submitted that year. Those seven investigators were under such extreme pressure to keep up with applications that they had little time to conduct their own research or maintain their professional education. In addition to new drug applications and complaints about misleading advertising, the same investigators also handled supplemental applications such as the one Searle had submitted for Enovid.

The Enovid application wound up on the desk of one of the part-timers, Pasquale DeFelice, a thirty-four-year-old obstetrician-gynecologist who was still completing his residency at Georgetown Medical Center. DeFelice, a native of Connecticut, was not only Catholic, he was a young man on his way to becoming the father of ten children.

The future of the pill was in his hands.

It was a strange moment for DeFelice. He was neither a career bureaucrat nor an experienced doctor. Yet he had in his power now the chance to review the work of far more experienced medical men and make a decision on the merits of their findings that would dramatically alter the world of medicine.

Even though he was Catholic and would go on to have a large family of his own, DeFelice disagreed with the Church's ban on birth control. He thought the pope had erred in taking such a firm line on contraception. But those were only his personal views. When it came to his job at the FDA, his faith and personal opinions were irrelevant. The only thing that mattered, he said, was whether the medication was safe and effective. In the case of Enovid, he had the sense that safety and effectiveness mattered even more than usual, because if the pill were approved, as he said, "everybody and her sister would be taking it."

As he sifted through the evidence and weighed the consequences of his action, DeFelice did not come under outside pressure, as far as the records of his work and his subsequent interviews indicate. No one in the White House or Congress weighed in publicly or tried to influence his decision behind the scenes. On the contrary, most officials in government seemed happy to avoid this particular controversy. No one in the Catholic Church made any effort to lobby for the application to be squashed and no one at Planned Parenthood pulled strings to try to help it get approved. In 1959, it was still possible for a bureaucratic process with huge social, economic, and religious stakes to come off without outside interference.

But that was about to change.

In the late 1950s, John D. Rockefeller began lobbying the U.S. government to make family planning part of its foreign aid work. In 1959, a presidential committee on military assistance programs headed by General William H. Draper backed Rockefeller's plea, recommending that foreign countries receiving military aid from the United States be encouraged to undertake birth-control programs.

When Eisenhower appointed Draper, he thought he was getting a conservative committee leader. After all, Draper was an investment banker and a former economic advisor to the postwar commander in chief of U.S. forces in Europe. But it turned out the general had a radical streak and wasn't afraid to toss grenades. After World War II, he had become concerned with population growth. He feared that Japan would soon have more people than it could feed. Much to the president's surprise, he argued in his report that America needed to get directly involved in international family planning so that developing countries would not be vulnerable to "communist political and economic domination." America's Catholic bishops immediately issued a statement of opposition, complaining that federal funds should never be used to promote contraception. On the presidential campaign trail, John Kennedy issued a statement backing the bishops and opposing American aid for family planning. His rivals for the Democratic nomination attacked, accusing Kennedy of attempting to impose his religious values on a pluralistic nation.

Eventually, Eisenhower distanced himself from the report, saying, perhaps disingenuously, that the U.S. government ought not to interfere with the internal affairs of foreign countries—even though the U.S. government did so all the time. "I cannot imagine anything more emphatically a subject that is not a proper political or governmental activity or function or responsibility," the president said at a news conference. Later, Eisenhower would change his mind and advocate federal support for family planning, but the first round in a battle over the government's role in birth control—a battle that would continue throughout the twentieth century and well into the twenty-first—went to the opposition. Congress rejected the Draper report.

※

Searle officials had hoped that DeFelice might rubber-stamp Enovid. The drug had already been approved, after all. They were merely asking for permission to prescribe it for another purpose. But as days turned to weeks and weeks to months, their hopes faded. DeFelice took his job too seriously to rubber-stamp anything. After three years at the agency, he found some of its conduct worrisome. If a company submitted an application for water to treat arthritis, he complained, the water stood a fair chance of winning approval because it did no harm and investigators couldn't prove that it *didn't* help arthritis. When ridiculous but difficult-to-reject applications of that variety crossed his desk, DeFelice sometimes requested extensions in the approval process. It was one of the few tools he had at his disposal to squeeze more information out of the companies pitching new drugs. By taking more time and insisting the company provide additional information, he hoped to find flaws in the application or, on the other hand, gain confidence that the drugs deserved his approval. Now, even though Searle had filed the most voluminous application in the history of American medicine with its petition for Enovid, DeFelice was determined to be meticulous in his review.

"When a New Drug Application came in for the birth-control pills," he told the writer Loretta McLaughlin, "it was, needless to say, revolutionary for that indication! It was a whole new bag of beans. Everything else up to that time was a drug to treat a condition. Here, suddenly, was a pill to be used to treat a healthy person and for long-term use." It was also likely to garner more attention than most if approved. The FDA, he said, could ill afford to make the wrong decision. The agency, he said, "had to come out of the licensing absolutely clean. . . . We were in no hurry to put the FDA stamp of approval on it."

After keeping the drug company waiting for more than two months, in late September DeFelice sent Searle officials a letter saying he needed yet more time. Too many questions remained, he wrote. DeFelice looked beyond the number of menstrual cycles Searle provided and bore down on the number of actual women tested. For example, how did the company know their pill was safe for twenty-four months when only 130 women in the trials had taken the drug for a year or more? Wasn't it possible the drug might trigger premature menopause if used for long durations? What about cancer, he asked, or the impact on other glands, or the long-term effect on future pregnancies?

He wrote: "We seriously question the validity of the use of a progestational agent . . . for the inhibition of ovulation, a normal body function, when there are so many unanswered questions concerning the potential for harm. This is especially true in view of the lack of advantage over other presently available contraceptive agents."

DeFelice had hit on the key issue. The pill treated a normal biological process as if it were a disease, and that made it unlike any other drug he could think of. The bar for approval had to be higher than usual. Why should women take something even the slightest bit risky when they were not sick and when condoms and IUDs offered safe if not exactly elegant alternatives? The application, he said, was "incomplete and inadequate."

※

Officials at Searle were furious.

The FDA had already approved the pill for the inhibition of ovulation. Why didn't the questions arise then, during the New Drug Approval process? The inhibition of ovulation was not a new claim, one Searle official wrote back in an angry rebuttal to DeFelice; inhibition of ovulation was the "inherent action of the drug . . . clearly outlined in our N.D.A." The letter went on to say that Pincus and Rock had found "no single case" of premature menopause, cancer, or trouble with future pregnancies. What cause did DeFelice have for raising those concerns?

John Rock was angry, too. How could the FDA hand a decision as important as this to a kid? What qualified *him* to judge? Hadn't he, the legendary Dr. Rock, already given Enovid his stamp of approval?

On a bitterly cold day in December 1959, Rock, along with I. C.

Winter and another Searle official, traveled to Washington, D.C., where they intended to confront DeFelice. The FDA inspector's office was located in a temporary wooden building that had been constructed during World War I. The lobby was freezing and there were no chairs. Rock and the men from Searle stood for ninety minutes while DeFelice kept them waiting.

To DeFelice, Rock was a legend, "the light of the obstetrical world," he said. "Anything he said had to be listened to."

But while he respected Rock as much if not more than any doctor, that didn't mean DeFelice was prepared to cave in to him—even when Rock spoke to him in a manner clearly intended to intimidate. Asked to explain his decision on Enovid, DeFelice said that he had to be more careful than he would with other drugs because healthy women would be taking it for extended periods of time. What if it turned out to cause cancer?

"I don't know how much training you've had in female cancer, young man," Rock said, "but I've had considerable."

DeFelice acknowledged that he lacked Rock's experience, but he noted that no one, not even Rock, had enough experience with Enovid to fully understand what it might do to women's bodies.

Rock fired back, shifting the argument away from personal health to the bigger issue of population control: "If your garage is on fire," he told DeFelice, "you do not wait to see if your bucket has a hole in it before trying to throw water on the blaze."

But DeFelice was not buying it and would not be cowed. He was a safety inspector, after all, not a firefighter. His job was to avert catastrophes, not to stop them from spreading. He said he wanted more data. He suggested, among other things, that Searle do additional lab tests to see if their drug caused blood clotting. Given that Enovid created pseudo-pregnancies in women and pregnant women tended to get more blood clots than other women, the question seemed reasonable and important to DeFelice.

When the meeting ended, Rock was still angry but Searle officials were humbled. They had little choice except to do what they'd been told. As for DeFelice, he remained an admirer of John Rock's. "I've only met about three doctors in my entire life who I would trust with anything," he said years later. "Rock was one of them. He still is."

While Searle went back to work to collect more data, DeFelice did more than wait. In an unusual though not unprecedented move for an FDA inspector, he launched a small research project of his own. DeFelice knew that thousands—and possibly even hundreds of thousands—of women were already using Enovid for birth control. He also knew that some of those women had found a way to get around the fact that their prescriptions were good only for three or four months. Many of them, when their initial prescriptions expired, simply made appointments to see new doctors and get new prescriptions. Such behavior might have been risky for the patients, but for DeFelice it presented an opportunity. Thousands of doctors across the country were gaining experience with Enovid—more every day. It was safe to assume that most of the prescribing doctors knew exactly which patients were taking the pill for birth control and how long they'd been taking it. If the doctors were doing their jobs well, they were probably gathering a great deal of information about the effects of this new medication. If side effects or health hazards had arisen in any significant numbers, the doctors likely would have heard about them. So DeFelice decided to contact some of those doctors and see what he could learn. He sent a questionnaire to sixty-one obstetrician-gynecologists at some of the leading medical schools in the United States, asking them a series of questions about their experiences with Enovid. Were the side effects serious enough to proscribe use of the product? Was long-term fertility affected? Were there any signs of cancer risk? Premature menopause? Were there any abnormalities among children born to women who used the medicine? And finally, he asked:

Do you think Enovid or another product like it ought to be made available for contraceptive purposes?

If the doctors prescribing the drug considered it safe, DeFelice would be more inclined to approve. If they didn't, if they thought the side effects were too severe, he would be inclined to reject.

Perhaps to prove to Searle that he was objective, DeFelice included Rock and Pincus (despite the fact that Pincus was not a physician) among his sixty-one experts, even though DeFelice already knew what both of them were going to say.

※

One of the doctors to whom DeFelice wrote was Edward Tyler, a former joke writer for Groucho Marx, who by 1959 was running the Planned Parenthood birth-control clinic in Los Angeles. Tyler had been conducting his own trials in Los Angeles, giving Enovid and a similar drug being developed by Parke-Davis to hundreds of patients. Outside of Puerto Rico, no one on earth had more experience administering birth-control pills than Dr. Tyler. His opinion meant a great deal to DeFelice, in part because Tyler, unlike Rock and Pincus, had no stake in the decision. He was not a consultant to Searle and he had not helped Pincus with the clinical trials in Puerto Rico.

Edward Tyler was administering the pill widely in his Los Angeles office, but he had reservations about its use. Too many of his patients, he said, experienced weight gain, fluid retention, and abnormal bleeding. At a 1958 meeting of the American Medical Association, Tyler had reported that more than two-thirds of his patients who had tried the pill had given it up, mostly because of the side effects. In his follow-up exams with women, Tyler detected symptoms of early menopause in two women. Most patients regained a normal menstrual rhythm after two months off the hormones, he reported, but he still feared that the pill might cause permanent change in the women's uteruses. He was also worried that female newborns of pregnant women who had been treated with the pill "not uncommonly" had adhesions of the labia and hypertrophy of the clitoris.

In 1958, Searle had invited Tyler and other doctors familiar with the new birth-control compounds to a meeting at the company's headquarters in Skokie. At that time, Tyler had expressed serious doubts about the wisdom of putting Enovid on the market. Now, in the early part of 1960, he would have a critical vote. FDA officials, after reading his written responses to their questionnaire, requested a personal interview.

Tyler still had concerns about the new oral contraception, but he also had a broader perspective. He was, after all, the head of one of Planned Parenthood's largest clinics. He knew what Margaret Sanger had first come to know back in the early part of the twentieth century: that the physical and emotional toll of nausea and bloating were nothing compared to the physical and emotional toll of an unwanted pregnancy, and that even the long-term risk of cancer might pale compared to the dangers of pregnancy or abortion for some women. The birth-control pill was still unproven, and it was far from perfect, but it was incredibly effective when it came to preventing pregnancy. That had to be weighed in his decision.

Tyler directly addressed the question that DeFelice had identified as so critical. How should the agency assess risk for a medication not designed to cure disease or relieve pain? As long as condoms and IUDs were an option, did it make sense for women to take such a chance?

If the FDA had assigned a woman to evaluate the pill in 1959, the agency's approach might have been different. A female investigator might have asked different questions. She might have surveyed patients instead of doctors, for example. But Tyler may have been the next best thing. He had seen firsthand the pill's effects—the good and the bad, the comfort and the pain. He was also clever enough to do some simple calculations. Diaphragms, condoms, and IUDs did not cause nausea or other side effects, but they did cause pregnancy because they had such high failure rates, and pregnancy came with its own long list of serious side effects, including preeclampsia, diabetes, hypertension, and heart attack. To analyze the risks of the birth-control pill effectively, one had to factor in the complications for women *not* using it.

After weighing all those factors, Tyler concluded that unwanted pregnancies would do more harm than the pill. He urged the FDA to approve it.

THIRTY-THREE

The Climax

IT HAD BEEN almost ten years since Gregory Pincus met Margaret Sanger on a winter night in an apartment high above Manhattan. Their chances of success at the time had seemed remote—almost ridiculously so. Sanger had been searching so long for her magic pill that even such a determined woman as she must have doubted at times whether she would succeed. Pincus had been a fringe player with nothing to lose. But there had been something durable and determined in each of them that kept them going through the years and the setbacks. Now, one way or another, their quest would come to an end. If the FDA rejected their application, there was no telling if they would get another chance. A Catholic might soon be in the White House. The FDA might tighten the rules surrounding drug approvals. G. D. Searle might get cold feet. Katharine McCormick might move on to another cause, or die and leave her money to others. Anything could happen.

And so Sanger and Pincus waited, and while they did, thalidomide began to reach customers in the United States. Richardson-Merrell, Inc., had already distributed more than two and a half million doses of its new sleeping pill—known by its trade name, Kevadon—to doctors across the country. Officials at the company hoped to get twenty thousand patients started on Kevadon and use the data from their experience in support of their pending application for federal approval. This drug trial would be one of the biggest in American history, much bigger than the trial for the birth-control pill. Richardson-Merrell was confident it would win FDA approval and thalidomide would soon be as popular in the United States as it was in Europe.

While Sanger and Pincus waited, Hugh Hefner opened the first Playboy Club, a so-called Disneyland for adults, where the famous Bunnies strolled in their electric-blue and kelly-green costumes, each breast looking to Norman Mailer "like the big bullet on the front bumper of a Cadillac," with little white tails bouncing on the ends of their bottoms. Within two years, the clubs would have three hundred thousand members.

In February 1960, while the FDA continued to review the application for Enovid, results of a Gallup poll showed nearly three out of four people believed birth control should be made available to anyone who wanted it.

Two months later, Leo Koch, a biology professor at the University of Illinois, was fired after writing a letter to the campus newspaper proposing a new approach to sex. "With modern contraceptives and medical advice readily available at the nearest drugstore, or at least from a family physician," he said, "there is no valid reason that sexual intercourse should not be condoned among those sufficiently mature to engage in it without social consequences and without violating their own codes of morality and ethics." His firing set off massive protests on campus, but the university refused to rehire him.

At the same time, an organization of respected judges, lawyers, and law professors in the United States known as the American Law Institute worked on one of the most ambitious projects in American legal history: an attempt to reform the nation's criminal codes, which varied from state to state and in many cases were so poorly written and organized that it was difficult to say what they meant or how they should be enforced. The judges and lawyers created a Model Penal Code that recommended to legislatures around the country how they might improve and standardize their own codes. One recommendation: Abortions should be legal when the pregnancy resulted from rape or when the baby was likely to be severely disabled.

It was part of a trend. The federal Comstock law and the "little Comstock laws" imposed by state governments were falling away. For most of American history, society and the law had seen womanhood and motherhood as virtually one and the same. Now, finally, that was changing, just as Margaret Sanger had always thought it would.

Still, she and Pincus waited. In those early days of 1960, Sanger wrote a letter to the *New York Times* criticizing President Eisenhower for knuckling under the pressure of the Catholic Church in refusing to support the Draper report. "Today our United States Government needs to listen to the voice of a majority of its voters," she wrote. "Birth control and contraceptive practices are a medically recognized part of current ethical healing." On the same morning Sanger's letter appeared, Senator Kennedy appeared on NBC's *Meet the Press*, where he was asked what he would do as a Catholic and as president if India asked for help from the U.S. government in controlling its population. It was yet another indication that Sanger's crusade was not so radical anymore. Once, it would have been unimaginable for a presidential candidate to be asked his views on birth control. Kennedy diplomatically answered that by putting such a controversial request to Congress "you'd get neither birth control nor foreign aid." The better approach, he said, would be to increase general economic assistance to India and let leaders of that country decide how best to spend it.

Sanger, not satisfied, followed up with a personal letter to Kennedy in which she tried to pin him down. She urged him to forget India and show leadership closer to home by speaking out against a Massachusetts law that limited women's access to contraception. "You are young," she wrote. "You come from a splendid, well cared for family, and it would be a great honor to you to raise your voice on behalf of the families of your church against this outrageous legislation." There is no record of Kennedy ever having responded.

Meanwhile, Pincus asked McCormick to approve a three-year budget for ongoing work on Enovid, including more safety tests. Whether or not the FDA approved the pill for contraception, much work remained to be done. In addition to his research on the pill, Pincus also wanted to conduct further experiments on a biological contraceptive for men. He had begun testing progestins on men in 1957, but he found that the hormones reduced the libidos and potency of his subjects, most of whom had been patients at the Worcester State Hospital. He thought it might be worth trying again with lower doses, and McCormick was encouraging him to do so. She also wanted him to see if Enovid would remain effective at lower doses. At ten dollars a month, it was simply too expensive, and she was eager to bring down the cost. At the same time, however, she did not intend to fund Pincus's research indefinitely. She had already committed $152,000 to Pincus ($1.2 million in today's dollars) for the first part of 1960, and she wanted to see what the FDA would decide on Enovid before giving him more.

While the FDA continued to process responses in its poll of physicians, the Family Planning Association of England was quietly testing the birth-control drug in London and Devon. The doctors leading those experiments concluded that the pill, while highly effective, produced too many side effects to be used by the general public.

In his Palm Sunday homily in 1960, Pope John XXIII reminded Catholics yet again where the Church stood on birth control as he urged parents to have many children. "Don't be afraid of the number of your sons and daughters," he said. "On the contrary, ask Divine Providence for them so that you can rear and educate them . . . to the glory of your fatherland here on earth and of that one in heaven." At the same time, leaders in the Vatican were growing concerned that other Christian churches were abandoning the ban on contraception, including the Anglican, Lutheran, and Calvinist communions, bodies whose general theology were not far removed from Catholic thought. Leading theologians like Reinhold Niebuhr, Karl Barth, and Emil Brunner were calling for a relaxation of the ban on contraception. Day by day, the Catholic Church was growing more isolated.

By the time of the pope's exhortation, four months had passed since John Rock's meeting with Pasquale DeFelice. The FDA still had not reached a decision. If the agency refused to grant Searle the right to sell Enovid for birth control, Rock wrote in April, "I am prepared to go to war with them."

※

By early May, the results were in from the FDA's survey of doctors. Of the sixty-one physicians responding to the agency's questionnaire, twenty-six recommended that the FDA approve Enovid for birth control; fourteen said they did not have enough experience with the pill to reach any conclusion; and twenty-one said they believed the FDA should deny its approval.

It was hardly an overwhelming show of support.

At least two of the doctors voting for denial cited religious beliefs. Some said they believed the drug was too expensive at fifty cents a dose, and others insisted that more testing should be required to prove the drug's long-term safety.

In an internal FDA memo, the agency's medical director, William H. Kessenich, acknowledged that the evidence on safety was thin. Most women had used the drug for no more than three or four months, and none of the women had used it for more than three years. "Only 66 patients have continued medication for 24 cycles or more," he wrote.

Kessenich also noted that the FDA's decision was likely to raise "possible objections from some quarters," an obvious reference to the Catholic Church.

But he also had to consider the agency's mandate. The FDA's job was to establish whether a drug worked. If it worked and there were no obvious indications that it caused harm, the agency had to approve. Even the doctors who urged the FDA to reject Enovid as contraception admitted that there were no signs of serious side effects "as far as they could tell."

That left the agency in a difficult spot. The bureaucrats at the FDA couldn't reject the pill on religious, political, or moral grounds. They couldn't say it didn't work. Nor could they say it caused harm. Uncertainty swirled around any new product. The only thing that was different about this one was that it happened to be associated with sex. For the FDA to reject the birth-control pill, it would have to have a solid scientific reason, and so far none had emerged.

※

On April 7, 1960, the phone on I. C. Winter's desk at Searle corporate headquarters rang.

Winter answered, and Pasquale DeFelice greeted him.

DeFelice gave him the news. In reviewing Searle's data, the FDA found no serious threat of side effects and no evident threat to overall patient health. What's more, there was no disputing that the drug worked effectively in preventing pregnancy. DeFelice said the agency had agreed to approve Enovid as an oral contraceptive.

There was only one caveat: The FDA wanted patients to limit their use of Enovid to two years in case the drug caused long-term side effects that had not come to light during the limited clinical trials. Those trials would continue, and if Searle wanted approval of the same pill at lower doses, they would have to reapply and provide additional scientific data.

The men discussed a few other matters. Searle would need to make minor changes in the text of its manual for doctors, and it would need to submit all of its labeling and advertising to the FDA before Enovid went on sale. But those were routine pieces of business. Searle officials wrote and mailed a memo the same day agreeing to every one of the FDA's conditions.

Just like that, it was done.

※

On May 9, 1960, the FDA issued its announcement, and newspapers across the country spread the news.

"U.S. APPROVES PILL FOR BIRTH CONTROL," read the headline in the *New York Times*. The article, only 136 words long, appeared on page seventy-five of the Tuesday, May 10, edition. "Approval was based on the question of safety," associate FDA commissioner John L. Harvey said in his official announcement. "We had no choice as to the morality that might be involved."

In the days ahead, dozens of other newspapers around the country carried the story. In most cities and towns, it did not make front-page news; newspapers were edited mostly by men and few anticipated the impact this development would soon have. The Associated Press account, which was the most widely published, began this way: "The federal government for the first time has approved a pill as safe for birth control." An official from Searle told the AP that the pill's function was "to interfere with the production of the ova in the same way nature does after a woman becomes pregnant." It was just the beginning of the company's work to put women at ease by stressing how the pill imitated nature. Soon John Rock would present the same argument to officials at the Vatican.

※

For Pincus, there was no time to celebrate. His daughter remembered no reaction from her father to the news. His personal and professional letters offer not the least hint that this was a momentous occasion. For him, the FDA's announcement on May 9 was not a finish line but a turn in the course. He made plans to return to Puerto Rico to expand testing of the smaller doses of his pill. If they worked, as he thought they probably would, the smaller doses might accomplish two important goals: reducing side effects and bringing down prices—and those two results in turn would help make the pill more popular.

Pincus might have had more reason to celebrate if he had maintained a financial stake in the development of Enovid. But he had never pursued a patent for his pill. Perhaps he had second thoughts about his decision now that it was reaching the market, but if he did, those concerns never materialized in his letters or in the minutes of his meetings with the Worcester Foundation's board of directors. Neither did any of his relatives or friends ever hear him complain about a missed opportunity to become wealthy. Through the years as the pill was in development, officials at Planned Parenthood and the Worcester Foundation had raised questions about patent rights. They had recognized, wisely, that if Pincus succeeded in making a birth-control pill and getting it approved, the pill had the potential to generate millions of dollars in profit. But all along, Pincus had discouraged such discussion. As far as he was concerned, he had not invented anything. His view was that his idea—like that of Jonas Salk and so many other scientists—had been built on the ideas and contributions of countless others. The pill was nothing new, according to this line of logic; it was merely nature refined and an extension of the body's own functions.

Salk had been widely quoted for his response after releasing his polio vaccine when Edward R. Murrow had asked him, "Who owns the patent on this vaccine?"

"Well, the people, I would say," Salk had answered. "There is no patent. Could you patent the sun?"

Pincus had almost certainly heard Salk's magnanimous response, and in the years ahead when the pill became enormously profitable, he would make similar comments, suggesting that there had really been nothing for him to patent. "Do not forget that we have never even made a pill at the foundation," he said. "All we did was to perfect the formula."

That was only partially true, of course. In reality, Pincus knew all along that Searle had developed the compound at the heart of Enovid. He also knew that other companies were working on copies of the pill and that each compound was slightly different. He might have tried to file his own patent once he perfected the mix of estrogen and progesterone in the pill, but such a move would have been difficult given how closely he'd been working with Searle. Pincus made the decision early in his career that science mattered more than money. He needed enough money to operate the Worcester Foundation and pay his associates, and he was aggressive about getting that money, but he never showed strong interest in the profit-making prospects of the work he and his fellow scientists did. In many ways he remained a purist, the same romantic spirit who, as a young man, had written sappy love poems and dreamed of making the world a better place.

On the other hand, he was not entirely altruistic. Searle helped pay his salary and provided generous support for the Worcester Foundation. Without its backing, and in particular without the free supplies of progesterone and progestins, his work might never have led to the pill. Pincus was also a Searle shareholder, and if the company's stock prices rose he stood to benefit, although even with the FDA's approval there was no immediate spike in the stock's price.

※

Once Enovid had been approved, Searle sent its army of sales representatives—or "detail men," as they were referred to in the pharmaceutical industry—out to the field to meet with doctors and urge them to begin prescribing the new pill. Before the 1950s, prescription drugs had been for the sick. But now new products were appearing that did not necessarily cure ailments so much as reduce the risk factors for future events such as heart disease or heart attacks. Doctors would decide which patients needed these drugs, which meant the detail men (and, yes, virtually all of them were men in 1960) needed to impress upon the doctors that their drugs should be prescribed liberally. Drug companies sometimes overwhelmed doctors with mailings on new products, but the detail men were expected to cut through the clutter and tell physicians what they needed to hear. Doctors, while skeptical, relied on the representatives as an important source of information on new drugs. For Enovid, Searle produced a twelve-page brochure with detailed information on clinical studies, toxicology reports, and animal-test results. In an in-house newsletter, Searle urged its detail men to "weed out all the negative points and convince doctors to get patients started on Enovid TODAY." The subjects of cancer, nausea, and religion were best avoided, the newsletter said. A better way to persuade the doctors was to remind them that patients taking the pill could be examined every month if the doctor or patient wished it. As primary-care physicians became more involved in family planning, their practices would grow. Women in perfect health would come to see them routinely. That, of course, meant money for the doctors as well as for Searle.

The detail men often handed out small presents to the doctors—notebooks, pens, and other trinkets intended to remind them that G. D. Searle had paid a visit. For Enovid, the representatives presented a special gift: a plastic paperweight painted gold. On the front was a three-dimensional representation of a naked, buxom woman, her arms breaking free from a set of heavy chains, her head tilted skyward. On the back was printed this message:

UNFETTERED

From the beginning, woman has been a vassal to the temporal demands, and frequently the aberrations, of the

cyclic mechanism of her reproductive system. Now, to a degree heretofore unknown, she is permitted normalization, enhancement, or suspension of cyclic function and procreative potential. ENOVID—the first comprehensive regulator of female cyclic function—is here symbolized in an illustration from ancient Greek mythology: Andromeda freed from her chains.

Almost immediately after winning FDA approval for the birth-control pill, Searle went back to the agency and asked for additional approval to sell Enovid in smaller doses. The FDA stalled, much to the drug company's frustration, but eventually gave its okay.

For Searle and for doctors, Enovid's arrival presented an opportunity to expand business and dramatically boost income. In 1957, the majority of family practitioners considered contraceptive counseling outside the scope of their jobs. The introduction of the pill quickly changed that. Even Catholic doctors felt compelled to write prescriptions when their patients requested it. After all, it was federally approved, and doctors who did hold out—either for religious reasons or because they feared the long-term side effects—found that they lost patients.

Searle took a conservative approach to marketing the pill in the first days after receiving FDA approval. The company focused its efforts primarily on doctors, not the public, in part because officials did not want to stir controversy and in part because they recognized that doctors would do most of the advertising for Enovid once they had the chance to see how it worked and how it generated new business. The market for Enovid grew steadily and quietly. Within four years, Searle's sales would increase 135 percent to $87 million, with a 38 percent return on equity for stockholders.

This is when the pill became widely known as The Pill, perhaps the only product in American history so powerful that it needed no name. Women went to their doctors and said they wanted it. They wanted The Pill. Some of them might still have been uncomfortable talking about birth control. Others might have been unsure of its brand name. But The Pill was The Pill because it was the only one that mattered, the one everyone was talking about, the one they needed.

※

Now that they'd done it, now that they'd finally created the thing that for so long had seemed an impossible dream, Sanger wrote to McCormick, jokingly asking what she intended to do for her next act.

Sanger was eighty. McCormick was eighty-four.

McCormick's reply was no joke. She wrote:

What I am busy over is as follows:

1. *Keeping on with the five branches of the Worcester Foundation and oral contraceptive work, namely, a) Dr. Rock's clinical tests on his patients; b) the Puerto Rico and Haiti field tests on women; c) clinical tests at the Worcester State Hospital, with intensive laboratory testing and study of long-term effects; d) laboratory research at WFEB to perfect Enovid.*
2. *Providing Shrewsbury housing for the fifteen post-doctoral students [working for Pincus].*
3. *First plans for the women's dormitory at MIT. I am particularly happy to be able to provide a dormitory on the Tech campus for women students there. This has been my ambition for many years, but it had to await the oral contraceptive for birth control.*

McCormick was indeed proud of her accomplishment, but she was also beginning to realize that the pill might not do as much to stem overpopulation as she had hoped. The high price of the medicine would prevent it from reaching many of the countries where it was needed most. That's why it was so critical that Pincus make improvements. Even if he did make the pill more effective and affordable, however, McCormick recognized that politics in some countries still might limit its reach. The best way to fight population growth, she admitted, might be to have more men undergo vasectomies. But even she didn't have enough money to make that happen.

Sanger, meanwhile, was struggling to kick her addiction to painkillers and to cut her alcohol intake. Though she still enjoyed periods of lucidity, she was showing signs of senile dementia. She managed to make headlines again in the summer of 1960 when she told reporters that if John Kennedy were elected president she intended to leave the country. "No one will really miss you," read one of the many angry responses she received.

"My heart aches for her all the time," one of Sanger's closest friends wrote in a letter to Sanger's son Stuart, who would soon ask a judge to rule his mother incompetent. A woman who had fought a thousand fights and won more than her fair share of them was, as a close friend described her, finally surrendering to "the ignominy of a rudderless drifting towards death."

Without Sanger to push them, Planned Parenthood did not embrace the pill right away. In fact, within weeks of the FDA's approval of Enovid, Planned Parenthood informed Pincus that it was cutting off all funding for his research; he didn't need the support of Planned Parenthood, officials said, now that "so much government and other money" was becoming available.

In the first years after Enovid's approval by the FDA, many Planned Parenthood officials continued to recommend IUDs, especially to poorer patients. These recommendations were in part because the price of Enovid remained high, in part because IUDs required no prescriptions, and in part because they were not sure poorer and less well-educated women could be trusted to take the pill every day. It was an old argument. For years, leaders of the population control movement had assumed that the poor lacked the proper motivation to make good use of contraception. But women of all income and education levels were now learning that contraception was more widely available and effective than ever before. They were beginning to understand that they didn't need to have seven or eight children and that once they controlled the timing of childbirth, they might begin to control all sorts of other things.

Sanger's crusade had begun when Woodrow Wilson was president and ended with Kennedy. It had begun when a woman devoting herself to anything but motherhood was by definition a radical.

Sanger and the pill did not quite ignite a sexual revolution, but they didn't have to. That fire was already burning by 1960; the pill only accelerated it. Yet in the grandest sense of all, Sanger's goal had been to make sex better—more pleasurable and loving—and by 1960, she had done just that.

※

As doctors and clinics embraced the pill and offered birth control to more women, John Rock led an aggressive public campaign to persuade ordinary Catholics and Catholic Church leaders to join the movement. He argued now, more loudly than ever, that the pill was a natural extension of the rhythm method and ought to be accepted. He appeared everywhere—in *Newsweek*, *Time*, *Reader's Digest*, the *Saturday Evening Post*, and on CBS and NBC. With his dignified air and impressive credentials, he gave the pill a kind of respectability that would have been impossible for Sanger or even for Pincus. FDA approval was important for women still uncertain whether to try the pill, but this seventy-year-old physician's sanction may have mattered even more, especially for Catholic women.

Rock wrote in *Good Housekeeping*: "The church hierarchy opposes use of the pill as immoral, but among communicants there is an increasing willingness to accept it. Close to half a million women are using the pill for contraceptive purposes. And it is hard for me to believe these women are all Protestants." Rock understood that most Catholic women were not waiting for an official pronouncement from the pope. They would decide for themselves. He made it his mission to help them choose correctly.

In the end, the Church disappointed him. Pope Paul VI ruled that the pill was just another form of artificial birth control and would not be permitted. But Rock never gave up hope. As he grew older and settled into retirement, he no longer attended daily Mass, but he always kept a crucifix above his desk, and he always believed that the next pope, or the one after that, would come around and see things his way.

※

Early in 1961, Pincus, along with his wife and researchers from the Worcester Foundation, returned to Puerto Rico to check on the continuing field trials. Pincus remained concerned about side effects, and he was eager to prove that his pill would work safely and effectively at lower doses, ensuring its long-term success.

Goody and Lizzie stayed at the Dorado Beach Hotel, in a ground-floor room with an ocean view, while most of the other scientists were housed at a less expensive hotel nearby.

One night before dinner, Goody and Lizzie invited everyone to their room for pre-dinner cocktails. The scene had the feeling of a victory party. The patio door was open and a cool ocean breeze blew in. Everyone was smoking and drinking and laughing. It had been less than a year since the FDA's approval of the pill, but already it was clear that these men and women had done something special, something perhaps bigger than they would ever do again, something that had fundamentally changed not only reproductive medicine but the lives of people everywhere. For one brief moment, while the party whirled on, Goody stepped out of the room. He strolled across the lanai and onto the grass that led to the beach, stopping for a moment to bend over and pluck a pink flower.

With his wife and colleagues watching from inside the hotel room, Pincus slid the flower behind his right ear and began to dance in the breeze to a song in his own head. Perhaps it was a song of his own making, the invention of a mind that had already given birth to something like a song, something that would set men and women free for generations to make love in cars on cold winter afternoons; in rowboats under moonlit skies; in corner offices late at night; in penthouses and dormitories; in houses, huts, and hotel rooms—in all the places where men wooed women or women wooed men, a spark was struck, and inhibition surrendered to desire. For generations to come there would be those who would hate Pincus, Sanger, McCormick, and Rock for what they had done, but just as surely there would be others in their debt, not only for the pleasure and passion the pill had supplied but also for the love, the opportunities, and the freedom it gave them.

Epilogue

WE NOW ACCEPT the pill as a part of life.

But looking back from a distance of more than half a century, it seems unbelievable that a group of brave, rebellious misfits—Sanger, Pincus, McCormick, and Rock—made such a radical breakthrough, and did it with no government funds and comparatively little corporate money. Indeed, there are countless ways that they might have failed. If Pincus had not been dismissed by Harvard and desperate to rebuild his reputation and career; if Sanger had not survived repeated heart attacks and maintained her ferocity even as she married into wealth; if Katharine McCormick's husband had not died and bequeathed her an immense fortune; if clinical trials had dragged on long enough for Americans to become aware of the thalidomide tragedy, the pill—described by Albert Mohler, president of the Southern Baptist Theological Seminary, as the most important development in history since the exile of Adam and Eve and referred to by the writer Mary Eberstadt as the "central fact" of our time—might never have been born.

※

In 1963, John Rock published *The Time Has Come*, a book he and his publisher described as "a challenge to solve the recurrent religious dispute over birth control." As a growing number of Catholic women ignored the pope and embraced the pill, Rock believed the Church might come around and give its approval. Debate raged everywhere, from local parishes in the United States to the highest levels of the

Vatican, from cocktail parties to network news. Rock became the most prominent face of the reform movement, and for a time in the early 1960s it appeared that his views might carry the day.

Soon after the release of his book, top officials in the Catholic Church invited the chairman of Planned Parenthood to meet with them at the Vatican. Another summit on birth control was held on the campus of the University of Notre Dame. In 1964, Pope Paul VI asked a committee of Church officials to reassess the Vatican's stance on contraception. Committee reports leaked to the *National Catholic Register* revealed that Rock's arguments were gaining favor with members of the committee, and that a majority of the committee members would recommend that the choice on birth control be left to women. But the pope, unimpressed, stalled for time, and while he stalled, theologians poked holes in Rock's argument. The pill was not a refinement of the rhythm method, they said. The rhythm method required abstinence during the fertile period, whereas the pill wiped out the fertile period. That made a big difference.

Finally, in 1968, Pope Paul VI issued the "Humanae Vitae" encyclical, stating clearly that all artificial methods of contraception violated the teachings of the Church. The pope labored over the final wording, no doubt aware that he risked coming across as out of touch with the thousands of Catholic women who had already made up their minds differently. He stressed the unifying characteristics of marriage, calling it the "wise institution of the Creator to realize in mankind his design of love." As for sex, the pope wrote, it must be "total"—a "special form of personal friendship in which husband and wife generously share everything . . . faithful and exclusive." But after the warm and fuzzy stuff, he got to the heart of the matter, saying that every act of conjugation must "remain open to the transmission of life." That meant the Church would not permit any action before or after sex intended to prevent procreation. Any conjugal act "deliberately rendered sterile," he wrote, "is intrinsically dishonest."

If sex for pleasure were permitted, Paul VI explained, moral standards would inevitably slide. Husbands would lose respect for their wives. Wives would lose respect for their husbands. Infidelity would flourish. The foundation of marriage would be weakened, perhaps catastrophically. Also, the pope said, if contraception became an accepted tool to control family size, oppressive governments might use it to coerce families to have fewer children.

The pope's declaration provided a pivotal moment, with some saying that the Church had missed its chance to adapt to the modern ethos and others saying it had taken an important stand for moral and religious values. Rock gained a small measure of vindication as hundreds of American theologians issued a statement asserting that the pope's decision was not an infallible teaching and Catholics were entitled to dissent.

In 1972, Rock retired from his practice, sold his house in Boston, and settled in a farmhouse in New Hampshire, where he swam in a stream behind the house, sipped martinis in front of the fireplace, and listened to John Philip Sousa marches on his record player. G. D. Searle paid him twelve thousand dollars a year for the rest of his life in what amounted to an unofficial pension and display of gratitude for his role in helping to bring Enovid to the world. He lived to the age of ninety-four, but not long enough to see the Church change its stance. It was one of the great disappointments in an almost charmed life.

"It frequently occurs to me, gosh, what a lucky guy I am," he said in one of the last interviews before his death in 1984. "I have everything I want. I take a dose of equanimity every twenty minutes. I will not be disturbed about things."

※

At least since the time of his trip to Japan in 1955, Pincus had been feeling ill. His stomach was frequently upset. In photos taken in the early 1960s he appears pale and drawn, the bags under his eyes heavier than usual. In August 1963, his personal physician ran a series of tests that showed Pincus's spleen was so swollen it filled the entire upper-left area of his abdomen. His prostate was enlarged, his white blood cell count was high, and his platelet count was almost off the charts. Bone marrow cancer was the doctor's best guess.

Pincus did not tell his colleagues about his diagnosis, but his friends and colleagues could see that he lacked his usual energy and no longer smiled as often. When rumors surfaced that he was ill, he denied them.

"I am healthier than I have been in many years," he told a colleague in 1966, and he continued to work.

The pill proved to be the great adventure of his life. He would often go out of his way to say that he never could have done it without McCormick, Sanger, Rock, M. C. Chang, Edris Rice-Wray, and many others. After the pill gained FDA approval in the United States, after sales began to rocket, and after Pincus began to read the letters from women for whom Enovid amounted to an answered prayer, his interests became more than scientific—he became a believer and an evangelist.

The executive secretary of the Planned Parenthood clinic in St. Paul, Minnesota, wrote to tell him that she had recently met a woman "who told us she had 'kissed' your picture (in our local newspaper)—she is so grateful to you, for this is the first year in her eight years of marriage that she has not been pregnant."

Some women complained about weight gain, others about the nausea. Many women experienced breast growth—to the extent that the sale of C-cup bras increased 50 percent from 1960 to 1969. The feminist Gloria Steinem switched from diaphragms to the pill in the early 1960s and wrote about it for *Esquire*. "For one thing," Steinem said, "it is more aesthetic than mechanical devices and, because it works chemically to prevent ovulation, it can be taken at a time completely removed from intercourse."

Pincus would dedicate the rest of his life to improving the pill and promoting it around the world, especially in Asia, where he traveled frequently. Perhaps his illness had something to do with it. He surely knew he would never attempt anything so ambitious again.

He made relatively little money on his invention—only his wages from Searle and the company stock he purchased. But he never complained because, since his childhood on the Jewish farmers' colony in New Jersey, he had never been compelled by money. He had led a comfortable life and done the work he'd always wanted to do. It had been important work, work that proved the power of his mind, work that had left its mark.

In 1961, four hundred thousand women were taking Enovid for birth control. A year later, that number tripled to 1.2 million. By 1964, Searle began selling Enovid-E with a dose of hormones of only 2.5 milligrams, reducing the cost for consumers to only $2.25 a month and reducing or eliminating the side effects for many. By 1965, more than 6.5 million American women were on the pill, making it the most popular form of birth control in the country. Around this time, as the pill became a bona fide phenomenon, some newspapers and magazines started spelling it with a capital *P*.

With success came scrutiny. Parents, teachers, and others fretted that Pincus's invention was spreading chaos as well as sexual pleasure. High school and college girls were talking about it, "and many are using it," an article in *U.S. News & World Report* noted in 1966. Cities were pushing the pill on welfare recipients, the same article reported. Journalists worried that sexual restraint would soon be a thing of the past, that sex would become informal, everyday, divorced of all romance, mystery, and taboo. "Will mating become casual and random—as among animals?" the magazine asked, echoing the same fear the Catholic Church had long raised. Already, there were reports of "sex clubs in high schools," wife swapping, and "housewives earning money as prostitutes—some with the knowledge and consent of their husbands."

College students began demanding that campus health centers dispense the new contraceptive. One gynecologist said he prescribed the pill "without qualms" to eight to ten coeds a month, noting, "I would rather be asked for the pills than for an abortion." In 1966, Pincus was asked by a reporter from *Candide* magazine to address accusations that in making the pill he had been "playing with the lives of women." He reminded the journalist that he had invented the pill for women at the request of a woman. In any case, he said, neither he nor the pill had played with anyone's life. Science was merely a tool. People used it as they wished. What's more, he said, the change was only beginning. Soon, he predicted, another drug would be available that women could take after sex to make sure they didn't get pregnant. He called it "The Next-Morning Pill," and he had already begun discussing it with a young French scientist named Étienne Baulieu, who would go on to use sex hormones to develop RU-486, the so-called abortion pill, which became available in France in 1988 and in the United States twelve years later. Pincus also predicted that infertile women would soon be able to use surrogates to carry babies for them. In short, he said, advances in reproductive biology would rapidly transform the way human beings were created. But that didn't mean scientists were playing with lives. They were merely exploring possibilities.

In 1965, Pincus published the book that summarized his life's work, *The Control of Fertility*, and dedicated it to "Mrs. Stanley McCormick because of her steadfast faith in scientific inquiry and her unswerving encouragement of human dignity." In 1966, he sold his accumulation of Searle stock, which had grown in value to about $25,000 ($180,000 in today's dollars). By the summer of 1967, he was living in terrible pain. His throat was constantly sore and his stomach ached.

In his final days, he sought to spend every possible minute with his wife. When Goody had to stay overnight at the hospital, Lizzie stayed, too.

On July 18, Pincus wrote to Katharine McCormick summarizing his latest research data and suggesting they arrange a meeting in the fall. He and McCormick both remained interested in a biological birth-control product for men, among other things. A month later, on August 22, 1967, he died at age sixty-four. He was buried beneath a tombstone that read "A GREAT AND KINDLY MAN."

※

If there was one problem with Pincus's invention, it was that even educated women sometimes had difficulty using it. Healthy young women were not accustomed to taking medicine every day. Sometimes they forgot, or they lost track of how many tablets they had taken since the start of a menstrual cycle. Nervous men found themselves reminding their wives and girlfriends, which led to friction as the men wondered if the women might secretly be trying to get pregnant and the women suspected that the men cared more about the women's sexual availability than their health. After one such marital spat, David P. Wagner of Geneva, Illinois, already a father of four, decided not to leave matters entirely in his wife Doris's hands. Wagner grabbed a piece of paper and put it on the dresser in their bedroom. On the paper, he wrote the days of the week. Then he placed one pill atop each day. When Doris swallowed a pill, the day of the week would be revealed and husband and wife both would have confirmation that she'd taken it.

"This did wonders for our relationship," Wagner said—until the paper fell one day and the pills scattered. Wagner, a product engineer for Illinois Tool Works, decided his wife needed a better container for her pills and began sketching a pillbox that would also function as a calendar. He took apart one of his children's toys and began working with a drill, tape, and some clear pieces of plastic.

In 1962, he applied for a patent on a circular pill dispenser and soon after paid a visit to the director of advertising at G. D. Searle in nearby Skokie. When Searle expressed no interest in his invention, Wagner sent a model to Ortho Pharmaceuticals, which was preparing to release its own birth-control pill. On February 1, 1963, when Ortho's contraceptive pill hit the market, it arrived not in a bottle but in a beautiful "Dialpak," shaped like a cross between a Frisbee and a UFO. It looked a lot like Wagner's invention. Ortho advertised the new packaging aggressively, hoping to distinguish its product from Searle's, which was far and away the industry leader.

When Wagner's patent was issued in 1964, more than a year after Ortho's Dialpak hit the market, Wagner and his lawyer informed the drug company that they intended to enforce the patent rights. Ortho paid Wagner a flat ten thousand dollars in exchange for the promise he wouldn't sue. Wagner then returned to Searle, asking the company to reconsider his invention, saying his original design was better than Ortho's and would help Searle eliminate any advantage Ortho might seize with its superior packaging. Searle still declined, declaring the dispenser nothing but a marketing gimmick. But when the drug company released Enovid-E in 1964, it came in yet another package closely resembling Wagner's design. Once more, Wagner and his lawyer complained. This time, the company agreed to pay him royalties. After legal fees, he wound up earning about $130,000, not only from Searle but from several other companies that adopted his design.

The pill's distinctive package helped make it one of the most easily recognized prescription drugs ever created. What's more, the pill now had a sleek, modern design that suited it perfectly and enhanced the product's popularity.

※

Katharine McCormick was too old and isolated in her Boston mansion to remotely comprehend how the pill was changing the lives of young women. But after pledging $1.5 million for the construction of a new women's dormitory at MIT, she did get to meet some of the young women who would be the beneficiaries of her generosity and foresight.

McCormick insisted on getting involved in the dorm's construction, just as she had insisted on guiding Pincus in his work on the pill. She wanted the new student housing to provide a healthy living environment that would help female students feel comfortable and secure on a campus still heavily dominated by men. Meetings on design were held in the parlor of her home. Though she suffered arthritis and dementia, she still dressed in proper business attire, including hat and gloves.

When the building opened, McCormick initiated weekly afternoon teas in the lobby, inviting students to socialize. She insisted that the women wear hats. Gloves were optional. This was in 1963, when such a request might have prompted ridicule, or at least snickers. But the young women of MIT made a game of it. They arrived in absurdly fancy hats and with baseball gloves and oven mitts on their hands. McCormick applauded their creativity.

A few years later, she agreed to pay for an additional wing on the dormitory. McCormick died on December 28, 1967, shortly after the wing's dedication. The building was named Stanley McCormick Hall.

※

"I knew I was right," Margaret Sanger told a journalist from the bed of a nursing home in 1963. "It was as simple as that. I knew I was right!"

If not for that conviction, Sanger might never have persisted in her quest for the birth-control pill. But that's not to say Sanger was entirely right. The pill did liberate women in many ways. It certainly gave them greater control over their sex lives and their family sizes. It undoubtedly opened up vast, new, and unimaginable opportunities to them. Yet when it came to sex, the pill had the opposite of

Sanger's desired effect for some women; it actually lowered their libidos. Sanger thought the pill would make married couples happier, but divorce rates have shot up since its advent. She also hoped the pill might lift women out of poverty and stop the world's rapid population growth. In fact, the pill has been far more popular and had greater impact among the affluent than the poor and has been far more widely used in developed countries than developing ones. In 1960, the global population stood at about three billion. Today it's about seven billion.

Even in Japan, where Pincus and Sanger worked so hard to generate enthusiasm and where abortion rates were among the highest in the world at the time, the government refused to approve the pill for decades for fear that it might promote promiscuity. Only in 1999, after the government approved Viagra, did Japanese officials relent and make the birth-control pill legal. Today advocates for birth control all over the world continue to wish for new contraceptives that might work more effectively in the developing world. But they face some of the same problems Sanger did before her meeting with Pincus in 1950, including a lack of enthusiasm among big pharmaceutical companies.

Sanger lived long enough to see that the pill was not entirely magic. But she also lived long enough to see birth control become a basic right of American citizens. In 1965, the U.S. Supreme Court ruled in *Griswold v. Connecticut* that the Bill of Rights includes a right to privacy and that the use of birth control was a private and protected act.

Sanger died eight months after the court's decision, a few days short of her eighty-seventh birthday. In a tribute, the Reverend Martin Luther King, Jr., called Sanger a woman who was "willing to accept scorn and abuse until the truth she saw was revealed to the millions." Jonas Salk wrote in tribute that "population growth, when uncontrolled, is like a disease; the cure must come from within the family of man. Margaret Sanger foresaw the danger and suggested a way."

Perhaps the most powerful comment on her life, however, came from the national Catholic weekly *Ave Maria*, which had excoriated her so many times in the past. In an editorial, the newspaper said Sanger's "vision was of a world in which all children would have from birth the opportunity to be fed and cared for, to be educated, to be loved. . . . Few of us are so hard-hearted that we fail to share her vision, whatever our reservations about her cause and means of birth-limitation."

※

In 1967, *Time* magazine put the pill on its cover, reporting that "in a mere six years it has changed and liberated the sex and family life of a large and still growing segment of the U.S. population: eventually, it promises to do the same for much of the world."

Attitudes toward sex were changing fast, thrillingly for some and horrifyingly for others. The pill didn't cause all these changes; it merely aided and abetted them. There were too many other forces at work for the pill to work alone. The bus boycott by African Americans in Montgomery, Alabama, launched a new era of activism. When the Civil Rights Act was introduced in Congress, feminists lobbied for the addition of an amendment prohibiting sex discrimination in employment. Soon after, Betty Friedan and other feminists founded the National Organization for Women (NOW). The movement against the war in Vietnam sparked a generation to rethink their methods of political and social rebellion and reimagine the power of the masses to effect change.

All of these social movements of the 1960s were about liberation, about challenging authority, about questioning convention. So-called Freedom Riders risked arrest to fight Southern segregation. Race riots erupted in the Watts neighborhood of Los Angeles. Antiwar protests disrupted college campuses. Women were in the thick of it, thanks in part to the pill. They postponed pregnancy, finished college, went to law school and medical school, applied for jobs, and took leading positions in government and the antiwar movement and the fight for equal rights. They also earned more money over the course of their careers.

In 1970, women comprised 10 percent of first-year law students and 4 percent of business school students; ten years later, those numbers jumped to 36 percent and 28 percent, respectively. And it wasn't just the women's movement making it happen. Harvard economist Claudia Goldin's research has shown that the pill had a direct effect. Women were more likely to enroll in graduate school and postpone marriage in states that lowered the age of consent for contraception from twenty-one to eighteen. The pill, Goldin concluded, lowered the cost of pursuing careers for women. No longer were they forced to sacrifice their social lives and prospects for marriage by choosing graduate school or ambitious career paths. In another study, economist Martha J. Bailey of the University of Michigan has shown that access to the pill boosted women's hourly wages by 8 percent and accounted for a whopping 30 percent of the convergence of the gender gap in earnings between 1990 and 2000.

In 1970, the median age at which college graduates married was about twenty-three. Five years later, it rose to about twenty-five and a half. When women did marry and did start families, the families were usually smaller than they had been a decade earlier. In 1963, 80 percent of non-Catholic college women wanted three or more children. A decade later that number dropped to 29 percent. In 1960, a typical American woman had 3.6 children. Two decades later, the number had dropped below two. In 1970, 80 percent of women with young children stayed home to care for the children and 20 percent worked. Today, those numbers have reversed.

※

The pill today remains one of the most widely prescribed drugs in the world. It is also one of the most widely examined. In the late

1960s and early 1970s, concerns arose about health risks associated with the pill, especially blood clots, and some leaders of the feminist movement began urging women to look for alternatives. Sales dipped briefly. Today, however, most research has concluded that the pill is not only safe but perhaps even beneficial in ways beyond contraception.

In 2010, British scientists released the results of a forty-year study, "Mortality Among Contraceptive Pill Users," that showed that women taking the birth-control pill were less likely than other women to die of heart disease, cancer, and other ailments. The study, which tracked forty-six thousand women, helped ease concerns about elevated risk of cancer or strokes. Women who took the pill were 12 percent less likely to die from any cause during the study. "Many women, especially those who used the first generation of oral contraceptives, are likely to be reassured by our results," said Philip Hannaford of the University of Aberdeen.

※

When he began his work testing progesterone on lab animals at the Worcester Foundation, when he was raising money by going door to door in the community where he lived, or when he was scrambling to assemble a few dozen subjects for experimentation by recruiting infertile patients of local gynecologists, Gregory Goodwin Pincus could not have dreamed that a clinical study would one day track tens of thousands of women over decades to check on the long-term effects of his pill.

Backed in part by the one million dollars bequeathed by McCormick, the Worcester Foundation operated into the 1970s, still focusing on women's health and conducting early research that would lead to the development of the breast-cancer drug Tamoxifen.

Pincus, Sanger, McCormick, and Rock no doubt would have been pleased that their legacy continued to produce important advances in women's health, but without their single-minded drive the Foundation fell upon hard times, and in the 1990s it merged with the University of Massachusetts Medical School.

Today the Foundation's grounds have been largely abandoned, with the exception of the Hoagland-Pincus Conference Center, which is still used by the university. A plaque there honors Hoagland and Pincus for their "advancement of knowledge . . . and betterment of mankind."

One day in the fall of 2011, Laura Pincus Bernard, Goody's daughter, walked through the empty halls of the ivy-covered building where her father once worked, where M. C. Chang once slept, where animals once mated or attempted to mate before giving their lives to science. The place was deserted except for a lone woman tapping a computer keyboard at a desk near the entrance.

Laura explained who she was and asked if she might look around. She climbed a narrow staircase to the attic, stepping carefully through a maze of dented file cabinets, old desks, chairs, and boxes filled with notebooks and loose sheets of paper containing the results of long-forgotten experiments. The desks were cheaply built—long plywood surfaces supported by metal drawers painted light shades of pink and avocado. Beakers and test tubes sat everywhere—in boxes, on countertops, and under venting hoods—their lids covered crudely in aluminum foil, as if having recently been sterilized and waiting for a scientist to return and use them again.

If it struck Laura as sad to see her father's foundation in such disrepair, she did not let on as she stepped across creaking floors through the detritus of the lab. Dust motes floated in the air. Outside, a school bus stopped and started again.

In a strange way, this old house served as a more fitting memorial than the conference center at the other end of the driveway. Even in its prime, the Worcester Foundation's headquarters had never impressed. The building, like the Foundation itself and the career of its founding scientist, had been a study in improvisation. Laura and others who had been there during the development of the pill knew what a close call the discovery had been—how success had sprung, more than anything, from the courage and conviction of the characters involved. That something so big and world changing had come from so humble a place seemed little short of a miracle.

Acknowledgments

I AM GRATEFUL TO so many people who devoted time, knowledge, and energy to this project. Gregory Pincus's daughter, Laura Pincus Bernard, shared family letters and photos, put me in touch with many of her father's associates, and accompanied me on a tour of the places where her father lived and worked. Rachel and Hart Achenbach shared their memories of Rachel's remarkable father, Dr. John Rock. Sue and Wes Dixon welcomed me to their home and told wonderful stories about Sue's father, Jack Searle. I was also fortunate to interview Isabelle Chang, wife of M. C. Chang.

Hundreds of other people gave generously of their time for interviews. Thanks in particular, and in no particular order, to Esther Katz; Cathy Moran Hajo; Dr. Henry Kirkendall, Jr.; Dr. Leonard Morse; Alex Sanger; Gloria Feldt; Larry Isaacson; Merry Maisel; Ronald Notkin; Andrew Pincus; David Pincus; Mike Pincus; Leo Latz, Jr.; Lex Lalli; Geoff Dutton; Evelyn Karet; Elizabeth Rubin; Erica Jong; Hugh Hefner; Dr. Edward E. Wallach; Ricardo Rosenkranz; Ellen More; Erica Jong; Kristine Reinhard; Tina Mercier; Neena Schwartz; Michael Moschos; Dr. Todd Hunter; Dr. Saul Lerner; Dr. Koji Yoshinaga; Dr. Prentiss C. Higgins; Dr. John McCracken; Dr. Nathan Kase; Judy McCann; Barbara Kupfer; Liza Gallardo; and Dr. Thoru Pederson.

I believe strongly in doing my own research. But scientists write more letters and keep better records than the ballplayers and gangsters I've written about in the past, which meant I needed extra hands and eyes to get through the materials stored in libraries and archives around the country. I am grateful to Lisa Applegate, Nick Bruno, Lauren Dickinson, Sonia Gomez, Chris Heidenrich, and Shane Zimmer for their research assistance. Special thanks go to Zimmer—researcher, editor, fact checker, spreadsheet builder, and friend—who has been with me almost from the start of this project. Ayako Mie helped dig up documents, photos, and newspaper clippings in Japan. For my research in Puerto Rico, I had assistance from Mike Soto, Anabellie Rivera, Daniel Epstein, Tyler Bridges, Marisol Lugo Juan, and Diana Rodriguez.

My friend Marci Bailey not only helped me search through library archives in Massachusetts, she also accompanied me on an eye-opening journey to Worcester, thoughtfully commented on my manuscript, and provided me a home away from home in Boston. My cousin, Dr. Jerry Avorn, read the manuscript and made helpful suggestions. Leslie Silverman, another cousin, pitched in with research. My brother, Matt Eig, and my friends Richard Babcock, Pat Byrne, Lou Carlozo, Mark Caro, James Finn Garner, Bob Kazel, Robert Kurson, Ron Jackson, and Jim Powers weighed in regularly with encouragement and advice. Bryan Gruley worked with me to map the path of the story and to make sure I stayed on course. Lori Rotskoff also read an early draft and helped me think more deeply about the book's themes. My friend and former teacher, Joseph Epstein, pushed me as he's been pushing me for thirty years to sharpen my writing. Other writer friends who pitched in along the way include Stephen Fried, Louise W. Knight, Gioia Diliberto, T. J. Stiles, Rachel Shteir, Jane Leavy, Rebecca Skloot, Chuck McCutcheon, Bob Spitz, Ben Kesling, and Charlie Newton. I am also thankful for good advice received from Linda Ginzel, Boaz Keysar, Sayuri Hayakawa, and Richard Thaler.

My friend Suzie Takacs of the Book Cellar in Chicago urged me to pursue this subject when I had doubts. The wonderful staff at

Unabridged Books in Chicago supplied me with loads of good reading. Thanks also to the top-notch staff at the Book Stall in Winnetka, Mitchell Kaplan at Books & Books in Miami, the Biographers International Organization, and the Tucson Festival of Books.

Jean Halberstam kindly granted me access to materials used by her late husband, David Halberstam, in his book *The Fifties*. Thanks to my friend Robert Solomon for introducing me to Ms. Halberstam. I'm also indebted to A. J. Baime for arranging my interview with Hugh Hefner. I want to thank Erna Buffie for sharing footage from her excellent documentary on the pill.

Kristen Meldi and Dr. Steven Sondheimer read my manuscript to make sure I got the science right, and Jack Cassidy checked it for everything else. Any mistakes that remain are my fault, not theirs.

I am also indebted to a number of librarians and archivists, none more so than Jeff Flannery at the Library of Congress, where I passed long hours immersed in the poems, letters, and scientific papers of Gregory Pincus. My thanks go out also to the staffs of the following institutions: the American Catholic History Research Center at the Catholic University of America; the Chicago History Museum; the Chicago Public Library (especially the John Merlo branch); the Clark University archives; the Countway Library of Medicine at Harvard University; the DePaul University library; the Kinsey Institute and the Lilly Library, both at the University of Indiana; the Lamar Soutter Library at the University of Massachusetts Medical School; the MIT Museum; the National Archives; the Sophia Smith Collection at Smith College; the University of Southern California libraries; the Wisconsin Historical Society; and the Worcester Historical Museum.

I must also thank the authors who explored the subject of birth control before I came around to it and put some of the building blocks of this story in place. The following writers took the time to offer their personal guidance: Annette B. Ramirez de Arellano, Laura

Briggs, Ellen Chesler, Esther Katz, Margaret Marsh, Gay Talese, and James Reed. In addition, Loretta McLaughlin and Leon Speroff—the biographers of John Rock and Gregory Pincus, respectively—met with me in person, provided access to their research materials, read my manuscript, and offered excellent suggestions.

I also benefited enormously from the work of the research team at the Margaret Sanger Papers Project at New York University, which has published a three-volume edition of Sanger's papers and a two-series microfilm edition of documents from the collections at Smith College.

This is my first book for W. W. Norton, and I'm grateful for the opportunity to work with such a talented and dedicated team. John Glusman is everything a writer could ask for in an editor: incisive, meticulous, and always pushing me to do my best. Thanks to Tara Powers for her scrupulous copyediting and to David High for his elegant design of the book's jacket. Also at Norton, thanks to Jonathan Baker, Louise Brockett, Steve Colca, Drake McFeely, Ingsu Liu, Jeannie Luciano, Nancy Palmquist, Jess Purcell, Don Rifkin, Bill Rusin, and Devon Zahn.

My agent, David Black, has been a steady believer in me—or my potential, anyway—for more than a decade. He and others at the David Black Agency, especially Antonella Iannarino and Sarah Smith, have been among my most indefatigable champions.

I've heard it said that writing is lonely work, but not for me. I've been encouraged, coddled, sustained, and entertained while working on this book for the past three years better than any writer could ever hope to be. I have my family to thank for that. My parents continue to urge me, as they have all my life, to work hard, follow my passions, and be creative. My daughters, Lillian and Lola, fill every day with laughter and inspire me to see the world through their wide-open eyes. Jeff Schams has been my weight-lifting partner, literally and figuratively, helping me stay strong and keep life's challenges in perspective. Finally, there's my wife, Jennifer Tescher, to whom this book is dedicated and to whom I owe thanks for everything—for her love, her wisdom, her endless support, not to mention her reading these pages early enough to make sure no one else would see the bad parts. We make a great team.

Notes

The narrative of this book is based on primary sources: thousands of letters and scientific reports; hundreds of scientific research papers; hundreds more newspaper and magazine articles; and interviews with more than one hundred people.

The majority of the documents were found at these archives:

Library of Congress, Washington, D.C. (LOC)
Sophia Smith Collection, Smith College, Northampton, MA (SSC)
University of Massachusetts Medical School Archive, Worcester, MA (UM)
Countway Library of Medicine, Harvard Medical School (CLM)
University of Southern California Libraries, Special Collections, Los Angeles, CA (USC)

CHAPTER ONE

1 *Winter, 1950*: Researchers, including this author, have spent countless long hours trying to determine the precise date of the first Pincus-Sanger meeting. In a 1953 letter to Sanger, Dr. Abraham Stone referred to a meeting with Sanger and Pincus at his home "two years ago." In *The Hormone Quest*, a book written with Pincus's cooperation, author Albert Q. Maisel sets the meeting on a "winter evening in 1950," which might have meant January, February, November, or December. But Pincus, in a letter to Al Raymond of G. D. Searle & Co. dated February 17, 1951, refers to a recent meeting with Stone in which they discussed a new research program on steroid contraceptives. Although it is clear from their comments that a meeting in New York occurred sometime that winter, personal diaries, calendars, and correspondence do not reveal a precise date. My reading of the evidence, along with an analysis of the travel schedules of the participants, suggests the historic encounter most likely took place in December 1950.

2 *"a street-fighting Jew"*: Dr. Enoch Callaway, telephone interview conducted by the author, March 2013.

3 *the request was denied*: Pincus to H. J. Muller, May 11 1942, Lilly Library, Indiana University, Bloomington, IN.

5 *"Let us see if we cannot begin to find our way"*: James R. Petersen, *The Century of Sex* (New York: Grove Press, 1999), p. 201.

6 *"Do you think that it would be possible . . . ?"*: "Creator of The Pill Talks to 'The Sun,'" *Sydney Sun*, January 9, 1967.

6 *"then start right away"*: Ibid.

6 *Chevrolet*: Laura Pincus Bernard, interview conducted by the author, October 2011.

7 *"This is just my cruising speed"*: Ibid.

7 *"linen cloth made to fit the glans"*: "Condom," *New York Times Magazine*, June 7, 2013, https://www.nytimes.com/packages/html/magazine/2013/innovations-issue/#/?part=condom (accessed February 19, 2014).

8 *"'the old ladies' home'"*: Robert C. Achorn, "Scientists at Shrewsbury Aim at Healthier Life," *Worcester Telegram*, September 3, 1947, p. 1.

8 *the paltry salary of $2,000 a year*: Isabelle Chang, telephone interview conducted by the author, July 2013.

8 *his room was at the YMCA*: M. C. Chang, "Recollections of 40 Years at the Worcester Foundation for Experimental Biology," *The Physiologist* 28, no. 5 (1985), p. 400.

8 *using Bunsen burners*: Isabelle Chang, telephone interview conducted by the author, July 2013.

8 *for one important experiment in 1947*: Chang, "Recollections of 40 Years at the Worcester Foundation," p. 401.

11 *$300 for miscellaneous supplies*: Gregory Pincus, March 16, 1951, Gregory Pincus Papers, LOC.

11 *"but I at once replied, 'Yes.'"*: Unpublished interview, *Candide*, Gregory Pincus Papers, LOC.

CHAPTER TWO

12 *"science and scientist continue to be governed by fear"*: Mary Roach, *Bonk* (New York: W. W. Norton, 2008), p. 12.

12 *textbooks . . . lacked entries*: Ibid.

15 *"attempting to lose their inhibitions"*: Malcolm Cowley, *Exile's Return: A Literary Odyssey of the 1920s* (New York: Penguin, 1994), p. 23.

16 *the key to curing neuroses*: Christopher Turner, *Adventures in the Orgasmatron* (New York: Farrar, Straus and Giroux, 2011), pp. 78–79.

16 *"heart ailments . . . excessive perspiration"*: Ibid., p. 80.

18 *"Fifties clothes were like armor"*: Brett Harvey, *The Fifties: A Women's Oral History* (New York: HarperCollins, 1993), p. xi.

19 *median age of marriage in 1950*: U.S. Bureau of Census report, September

15, 2004, http://www.census.gov/population/socdemo/hh-fam/tabMS-2.pdf (accessed February 18, 2014).

19 *"What's college?"*: Elizabeth Siegel Watkins, *On the Pill* (Baltimore, MD: Johns Hopkins University Press, 1998), p. 9.

20 *"Birth control would have been cold-blooded"*: Harvey, *The Fifties*, pp. 11–12.

20 *"I was terribly frightened about getting pregnant"*: Ibid., p. 12.

20 *Most American women . . . accepted the idea of birth control*: Watkins, *On the Pill*, p. 11.

20 *not a question of principle*: Birth Control Hearings before the Committee on the Judiciary, House of Representatives, 73rd Congress, 2d sess., on H.R. 5978, Jan. 18–19, 1934 (Washington, D.C., 1934), SSC.

CHAPTER THREE

22 *"tennis or chess"*: Bernard Asbell, *The Pill: A Biography of the Drug That Changed the World* (New York: Random House, 1995), p. 124.

22 *"Victory!"*: Unpublished interview, *Candide*, Gregory Pincus Papers, LOC.

22 *might get them more money*: Asbell, *The Pill*, p. 124.

23 *"'cunning device'"*: Gregory Pincus, *The Control of Fertility* (New York: Academic Press, 1965), p. 6.

23 *"consequences that are not apparent on the surface"*: Ibid., pp. 6–7.

23 *"ivory tower conception of research"*: Ibid., p. 7.

23 *the world in which they lived*: Ibid., p. 8.

24 *"though dull of mind"*: Matthew James Connelly, *Fatal Misconception: The Struggle to Control World Population* (Cambridge, MA: Belknap Press of Harvard University Press, 2008), p. 61.

25 *"more serious than the atomic bomb"*: "Creator of The Pill Talks to 'The Sun,'" *Sydney Sun*, January 9, 1967.

25 *wool-spinning machines, and electric clocks*: *Polk's Worcester City Directory* (Detroit, MI: R. L. Polk and Company, 1954), pp. 8–9.

26 *Wear-Well Trouser Co. and the Worcester Baking Company*: Worcester Foundation annual reports and internal reports, Worcester Foundation Papers, UM.

27 *"Since sleep escapes me"*: Undated letter, Gregory Pincus to Albert Raymond, Gregory Pincus Papers, LOC.

28 *"I want you to know"*: Ibid.

CHAPTER FOUR

29 *she wrote to a friend and supporter in 1939*: Margaret Sanger to Clarence Gamble, August 15, 1939, Margaret Sanger Papers, SSC.

29 *"Dear Mrs. Sanger . . ."*: Margaret Sanger Papers, SSC.

31 *revealing the true angels within*: "The Child Who Was Mother to a Woman," *The New Yorker*, April 11, 1927.

31 *His friends loved him and trusted him*: Ibid.

31 *"It was Father"*: Margaret Sanger, *My Fight for Birth Control* (New York: Farrar & Rhinehart, 1931), pp. 11–12.

31 *a constable barring the door*: "The Child Who Was Mother to a Woman," *The New Yorker*, April 11, 1927.

32 *where Ingersoll finally spoke*: Ibid.

32 *"the juvenile stamp of disapproval"*: Margaret Sanger, *The Autobiography of Margaret Sanger* (Mineola, NY: Dover, 2004), p. 20.

32 *"they were wrong"*: "The Child Who Was Mother to a Woman," *The New Yorker*, April 11, 1927.

32 *With financial support from her older sisters*: Ellen Chesler, *Woman of Valor* (New York: Simon & Schuster, 2007), p. 30.

32 *"I longed for romance"*: David M. Kennedy, *Birth Control in America* (New Haven, CT: Yale University Press, 1970), p. 5.

33 *marriage "akin to suicide"*: Sanger, *My Fight for Birth Control*, p. 31.

33 *"I was sick for two months"*: Sanger, *The Autobiography of Margaret Sanger*, p. 57.

33 *"Socialists, Trade Unionists, Anarchists"*: William B. Scott and Peter M. Rutkoff, *New York Modern: The Arts and the City* (Baltimore, MD: Johns Hopkins University Press, 1999), p. 81.

34 *"stands firmly by its roots"*: Journal entry, November 3–4, 1914, Margaret Sanger Papers, SSC.

34 *"an ardent propagandist for the joys of the flesh"*: Peter Engelman, *A History of the Birth Control Movement in America* (Santa Barbara, CA: ABC-CLIO, 2011), p. 31.

34 *found the conditions "almost beyond belief"*: Sanger, *My Fight for Birth Control*, pp. 46–48.

34 *below Fourteenth Street east of Broadway*: "New York Wards: Population and Density, 1800–1910," Demographia.com, http://www.demographia.com/db-nyc-ward1800.htm (accessed February 19, 2014).

35 *460 square feet each*: "Manhattan's Population Density, Past and Present," *New York Times*, March 1, 2012.

35 *population had increased 62 percent*: "New York Wards: Population and Density, 1800–1910," Demographia.com, http://www.demographia.com/db-nyc-ward1800.htm (accessed February 19, 2014).

35 *"Poor pale faced wretched wives"*: Margaret Sanger to Juliet Barrett Rublee, July 7, 1920, Margaret Sanger Papers, SSC.

35 *one-third of all pregnancies*: "The Question of Birth Control," *Harper's* magazine, December 1929, p. 40.

35 *"inserting slippery-elm sticks"*: Sanger, *My Fight for Birth Control*, p. 47.

36 *I would be heard*: Ibid., p. 56.

36 *lapsed into severe depression*: Chesler, *Woman of Valor*, p. 52.

36 *the maid would be there*: David Halberstam, *The Fifties* (New York: Random House, 1994), p. 283.

37 *7.04 in 1800 to 3.56 in 1900*: Daniel Scott Smith, "Family Limitation, Sexual Control, and Domestic Feminism in Victorian America," *Feminist Studies* 1, no. 3–4 (1973), pp. 40–57.

38 *24 percent for these devices*: *Controlling Reproduction*, ed. Andrea Tone (Wilmington, DE: Scholarly Resources, 2001), p. 75.

38 *"except total abstinence"*: Ibid., p. 81.

38 *plant-fiber tampon coated with honey . . . and swallowed poisons*: "Leeches, Lye and Spanish Fly," *New York Times*, January 22, 2013.

39 *disease of both body and soul*: Chesler, *Woman of Valor*, p. 52.

39 *"You are a world Lover"*: Bill Sanger to Margaret Sanger, February 6, 1914, Margaret Sanger Papers, SSC.

39 *"go-to-hell look"*: "The Aim," *The Woman Rebel*, March 1914.

CHAPTER FIVE

41 *"the severed head of Holofernes"*: "The Child Who Was Mother to a Woman," *The New Yorker*, April 11, 1927.

42 *"in the whole of his life"*: Arthur Calder-Marshall, *The Sage of Sex: A Life of Havelock Ellis* (New York: G. P. Putnam's Sons, 1959), pp. 197–98.

42 *"ever wonderful, ever lovely"*: Henry Havelock Ellis, *The New Spirit* (London: Walter Scott, 1890), p. 129.

42 *"It is wonderful enough"*: H. G. Wells, *The Secret Places of the Heart* (New York: MacMillan, 1922), p. 250.

43 *Comstock had masturbated so obsessively*: Gay Talese, *Thy Neighbor's Wife* (New York: Doubleday, 1980), p. 53.

44 *"any obscene, lewd, lascivious"*: Section 211 of the U.S. Criminal Code, http://books.google.com/books?id=6cUZAAAAYAAJ&pg=PA10381-IA2&lpg=PA10381-IA2&dq="any+obscene,+lewd,+or+lascivious":+Section+211+of+the+U.S.+Criminal+Code&source=bl&ots=_m3p115xFc&sig=0D4DBGx_m71oj1pbbHGiBPfsD5o&hl=en&sa=X&ei=grMQU6SiH6LWyQH-5YHwDQ&ved=0CCkQ6AEwAA#v=onepage&q="any%20obscene%2C%20lewd%2C%20or%20lascivious"%3A%20Section%20211%20of%20the%20U.S.%20Criminal%20Code&f=false (accessed February 28, 2014).

44 *"a weeder in God's garden"*: Talese, *Thy Neighbor's Wife*, p. 56.

44 *"You are all the world to me"*: Ellen Chesler, *Woman of Valor* (New York: Simon & Schuster, 2007), p. 106.

45 *Stuart, alone at boarding school*: Ibid., p. 107.

45 *"reflect, meditate and dream"*: Ibid.

45 *haunted by nightmares*: Ibid., p. 134.

46 *"Then we got a little nearer"*: Peter C. Engelman, *A History of the Birth Control Movement in America* (Santa Barbara, CA: Praeger, 2011), p. xviii.

47 *"wrists reddened" . . . "what scars Murray and Foley are nursing"*: "Mrs. Sanger Flays Mrs. Davis' Plans," *New York Tribune*, March 7, 1917; "Mrs. Sanger Free, Hailed as Heroine," *New York Tribune*, March 6, 1917.

48 *"constant tendency"*: T. R. Malthus, *An Essay on the Principle of Population* (Cambridge: Cambridge University Press, 1992), p. 14.

49 *"I wasn't doing my duty as a wife"*: Margaret Sanger, *Motherhood in Bondage* (Columbus: Ohio State University Press, 2000), p. 124.

49 *Sanger wrote in 1919*: Margaret Sanger, "Parent's Problem or a Woman's," *Birth Control Review* 3, no. 3 (1919), pp. 6–7.

50 *"The Church's attitude on birth control"*: "The Question of Birth Control," *Harper's Monthly Magazine*, December 1929.

50 *"the sacramental attitude"*: Ibid.

51 *There were conditions*: Lawrence Lader, "Margaret Sanger: Militant Pragmatist Visionary," *On the Issues* Magazine, Spring 1990, http://www.ontheissuesmagazine.com/1990spring/Spr90_Lader.php (accessed February 19, 2014).

51 *"retire with him to the garden of love"*: *Controlling Reproduction*, ed. Andrea Tone (Wilmington, DE: Scholarly Resources, 2001), p. 129.

51 *"the greatest adventure in my life"*: Lawrence Lader, "Margaret Sanger: Militant Pragmatist Visionary," *On the Issues* Magazine, Spring 1990.

52 *"234 clinics and 140 hospitals"*: Ibid., p. 134.

53 *weeding out of the "unfit"*: Ellen Chesler, *Woman of Valor: Margaret Sanger and the Birth Control Movement in America* (New York: Simon & Schuster, 2007), p. 195.

54 *in the backseats of their cars*: Jean H. Baker, *Margaret Sanger: A Life of Passion* (New York: Hill and Wang, 2011), p. 174.

55 *"To take life after its inception"*: "Archbishop Hayes on Birth Control," *New York Times*, December 18, 1921.

55 *"What he believes"*: Typed statement by Margaret Sanger, January 20, 1921, Margaret Sanger Papers, SSC.

55 *as Chesler wrote*: Chesler, *Woman of Valor*, p. 470.

56 *"There is no need to summarize"*: Andrea Tone, *Devices and Desires: A History of Contraceptives in America* (New York: Hill and Wang, 2001), p. 147.

57 *"a fine piece of research"*: Margaret Sanger to Katharine Dexter McCormick, January 8, 1937, Margaret Sanger Papers, SSC.

CHAPTER SIX

58 *at least one unwanted pregnancy*: Brett Harvey, *The Fifties: A Women's Oral History* (New York: HarperCollins, 1993), p. 92.

59 *"modern as tomorrow"*: Betty Millburn, "A Witty Friend and a Gracious Hostess," *Tucson Citizen*, *Arizona Daily Star* archives, undated.

59 *hobnobbed with the social elite*: Margaret Regan, "Margaret Sanger: Tucson's Irish Rebel," *Tucson Weekly*, March 11, 2004, http://www.tucsonweekly.com/tucson/margaret-sanger-tucsons-irish-rebel/Content?oid=1075512 (accessed February 20, 2014).

59 *some of the money*: Madeline Gray, *Margaret Sanger: A Biography of the Champion of Birth Control* (New York: Richard Marek Publishers, 1979), p. 292.

59 *Prescott S. Bush*: "Bush Family Planning," Margaret Sanger Papers Project newsletter, Winter 2006/2007, no. 44, New York University, http://www.nyu.edu/projects/sanger/newsletter/articles/bush_family_planning.htm (accessed February 20, 2014).

60 *in Sanger's words*: Katharine Dexter McCormick to Margaret Sanger, October 27, 1950, Margaret Sanger Papers, SSC.

60 *"Confidential Please"*: Margaret Sanger to Clarence Gamble, dated "Saturday/October 1942," Gamble Papers, CLM.

CHAPTER SEVEN

62 *"Our one duty"*: Pincus diaries, September 19, 1920, Gregory Pincus Papers, LOC.

62 *"Greatness is a spiritual condition"*: Pincus diaries, undated entry, Gregory Pincus Papers, LOC.

64 *"He was so handsome"*: Leon Speroff, M.D., *A Good Man: Gregory Goodwin Pincus* (Portland, OR: Arnica Publishing, 2009), p. 48.

64 *"How many nights I cried"*: Ibid.

65 *Rabbi Stephen Wise's Free Synagogue*: Alex Pincus, unpublished memoir, Gregory Pincus Papers, LOC.

65 *"took him into bed"*: Ibid.

65 *"for the same joy in the future"*: Pincus diaries, January 20, 1920, Gregory Pincus Papers, LOC.

65 *"the embodiment of all our ideals"*: Pincus diaries, March 7, 1920, Gregory Pincus Papers, LOC.

65 *"I have not heretofore strictly practiced"*: Pincus diaries, January 27, 1920, Gregory Pincus Papers, LOC.

65 *"the holiest passion on this earth"*: Pincus diaries, January 8, 1920, Gregory Pincus Papers, LOC.

65 *"affectionate being"*: Pincus diaries, March 21, 1920, Gregory Pincus Papers, LOC.

66 *"Values and standards which have been"*: Gregory Pincus to his mother, undated letter, Gregory Pincus Papers, LOC.

67 *washing dishes and waiting tables*: Pincus diaries, March 21, 1920, Gregory Pincus Papers, LOC.

67 *his family during vacations*: Ibid.

67 *she wrote in a memoir*: Elizabeth Pincus, unpublished memoir, family collection.

68 *"clear conscience and a big heart"*: Gregory Pincus to his mother, undated letter, Gregory Pincus Papers, LOC.

68 *Philip Morris cigarettes*: Speroff, *A Good Man*, p. 125.

68 *"Growing a penis"*: Geoff Dutton, interview conducted by the author, October 2011.

68 *"I'm a sexologist"*: Speroff, *A Good Man*, p. 74.

69 *"epicenter of American education"*: Richard Norton Smith, *The Harvard Century* (Cambridge, MA: Harvard University Press, 1998), p. 12.

70 *What form of birth control . . . no one knows*: Laura Pincus Bernard, interview conducted by the author, October 2011.

70 *"I wanted to take life in my hands"*: Philip J. Pauly, *Controlling Life: Jacques Loeb and the Engineering Ideal in Biology* (New York and Oxford: Oxford University Press, 1987), p. 102.

71 *"a very good probability of failure"*: Gregory Pincus, unpublished manuscript, Gregory Pincus Papers, LOC.

72 *"an arrogant bunch of brats"*: Hudson Hoagland, "Change, Chance and Challenge," unpublished memoir, UM.

72 *influential psychologists and behaviorists*: Pauly, *Controlling Life*, p. 191.

72 *Pincus's contract was only barely approved*: Ibid.

73 *apply his* in vitro *technique to humans*: Ibid.

73 *"At Harvard are two Bokanovskys"*: "The Week in Science," *New York Times*, May 13, 1934.

74 *"the tinkering of a biological Edison"*: Pauly, *Controlling Life*, p. 192.

74 *"bound to receive his meed of the praise"*: "BOTTLES AS MOTHERS," *New York Times*, April 21, 1935.

75 *"MANLESS WORLD?"*: "Manless World?" *Racine Journal-Times*, April 15, 1936.

75 *"Pfluger, 1863"*: Gregory Pincus, *The Eggs of Mammals* (New York: Macmillan, 1936), pp. 8–9.

76 *"The social implications of Dr. Pincus's"*: "Brave New World," *New York Times*, March 28, 1936.

77 *"In the huge Biological Laboratory"*: "No Father to Guide Them," *Collier's*, March 20, 1937.

78 *Hoagland drove to the drug store*: Enoch Callaway, *Asylum: A Mid-Century Madhouse and Its Lessons about Our Mentally Ill Today* (Westport, CT: Praeger, 2007), p. 18.

79 *"Knowing his brilliance"*: Hudson Hoagland, "Change, Chance and Challenge," unpublished memoir, UM.

80 *sprawled across the surrounding yard*: Laura Pincus Bernard, telephone interview conducted by the author, December 2012.

80 *Hoagland recalled*: "Biology Foundation Spawned in a Barn," *Worcester Telegram*, June 8, 1952.

80 *"The Clark work was reported"*: "Test Tube Ova Demonstrate Tendency of Reproduction," *Jefferson City Post-Tribune*, April 28, 1939.

80 *The AP corrected its error*: "Test Tube Furore is Result of Omitted 'Not'," *Ogden Standard-Examiner* (Utah), May 19, 1939.

81 *with encouraging preliminary results*: G. Pincus and H. Hoagland, "Effects on Industrial Production of the Administration of Pregnenolone to Factory Workers, I," *Psychosomatic Medicine* 7, no. 6 (1945), pp. 342–46.

81 *opening session of the conference*: Transcript of Joseph Gildzieher interview conducted by Leon Speroff, October 2007.

81 *the principal of John's high school*: Laura Pincus Bernard, interview conducted by the author, October 2011.

81 *room, board, and clothing*: Ibid.

82 *what an ordeal her home life had become*: Ibid.

83 *began emitting its contents into the groundwater*: "The Growth and Future of the Worcester Foundation: A Report to the Trustees," 1950, UM.

83 *how to pay*: Ibid.

83 *a portable table, and a keg of nails for a chair*: Transcript of Jackie Foss interview conducted by Leon Speroff, May 2007.

83 *"growth was either wise or necessary"*: "The Growth and Future of the Worcester Foundation: A Report to the Trustees," 1950, UM.

84 *always surrounded by towers of books*: Michael Moschos, telephone interview conducted by the author, June 2013.

84 *"It's like living in a madhouse"*: Alex Pincus, unpublished memoir, Gregory Pincus Papers, LOC.

85 *"The Serious Stinkers"*: Laura Pincus Bernard, interview conducted by the author, July 2013.

86 *$30,000*: Speroff, *A Good Man*, p. 117.

86 *paid for with Worcester Foundation money*: Laura Pincus Bernard, interview conducted by the author, July 2013.

86 *J&B or Johnny Walker*: Geoff Dutton, interview conducted by the author, October 2011.

86 *Philip Morris cigarette dangling*: Isabelle Chang, interview conducted by the author, July 2013.

86 *"You give her a test-tickle"*: Geoff Dutton, interview conducted by the author, October 2011.

86 *"she had this sort of demeanor"*: Transcript of Michael Bedford interview conducted by Leon Speroff, undated.

86 *until Lizzie relented*: Laura Pincus Bernard, interview conducted by the author, July 2013.

87 *"an important factor in Goody's life history"*: Alex Pincus, unpublished memoir, Gregory Pincus Papers, LOC.

87 *refused to get behind the wheel*: Laura Pincus Bernard, interview conducted by the author, July 2013.

87 *Mrs. Pincus could be so disagreeable*: Transcript of Jackie Foss interview conducted by Leon Speroff, May 2007.

87 *as soon as she woke*: David Halberstam, *The Fifties* (New York: Villard, 1993), p. 289.

88 *remembered her father sleeping*: Ibid.

88 *"a man who . . . was indestructible"*: Oscar Hechter, "Homage to Gregory Pincus," *Perspectives in Biology and Medicine* 11 (Spring 1968), p. 367.

88 *"Nobody dared tell him a lie"*: Isabelle Chang, interview conducted by the author, July 2013.

89 *Sheldon Segal*: Sheldon Segal, "Gregory Pincus, Father of the Pill," Population Reference Bureau, http://www.prb.org/Publications/Articles/289/GregoryPincusFatherofthePill.aspx (accessed February 18, 2014).

89 *"When you went to the Laurentian"*: Ibid.

89 *"like an emperor"*: Ibid.

CHAPTER EIGHT

90 *"Mrs. Stanley"*: Katharine Dexter McCormick to Margaret Sanger, October 27, 1950, Margaret Sanger Papers, SSC.

93 *masturbated publicly*: Armond Fields, *Katharine Dexter McCormick: Pioneer for Women's Rights* (Westport, CT: Praeger, 2003), p. 150.

94 *plotting their attack*: Ibid., p. 177.

95 *including three large trunks*: Ibid., p. 181.

95 *which was putting it mildly*: Ibid., p. 213.

96 *cost another $108,000 a year*: Obituary, *Santa Barbara News-Press*, January 20, 1947.
96 *"executrix of the estate"*: Fields, *Katharine Dexter McCormick*, p. 252.
96 *almost thirty-two thousand shares*: Ibid.
96 *her husband having died in 1943*: Ellen Chesler, *Woman of Valor: Margaret Sanger and the Birth Control Movement in America* (New York: Simon & Schuster, 2007), p. 399.
97 *long ago been transferred to her name*: Ibid.

CHAPTER NINE

98 *"feeling pretty desperate"*: Katharine Dexter McCormick to Margaret Sanger, January 22, 1952, Armond Fields Collection, USC.
99 *try more progesterone compounds*: "Report of Progress to: Planned Parenthood Federation of America, Inc.," January, 24, 1952, LOC.
99 *"to those who may help support it"*: William Vogt to Gregory Pincus, April 21, 1952, LOC.
99 *"evidently not been sold"*: James Reed, *From Private Vice to Public Virtue: The Birth Control Movement and American Society Since 1830* (New York: Basic Books, 1978), p. 341.
99 *on other business*: Katharine Dexter McCormick to Margaret Sanger, June 20, 1952, Margaret Sanger Papers, SSC.
100 *"As scientists and individuals"*: "General Summary and Results," Arden House Colloquium on Human Fertility, September 13–14, 1952, Gregory Pincus Papers, LOC.
100 *"I am rather surprised"*: Katharine Dexter McCormick to Margaret Sanger, October 1, 1952, Armond Fields Collection, USC.
101 *"It is pretty trying"*: Katharine Dexter McCormick to Margaret Sanger, March 15, 1953, Armond Fields Collection, USC.
101 *"we can go see Dr. Pincus"*: Margaret Sanger to Katharine Dexter McCormick, March 27, 1953, Armond Fields Collection, USC.

CHAPTER TEN

102 *"He wasn't afraid to go out on a limb"*: Seymour Lieberman, telephone interview conducted by the author, October 2011.
104 *"Don't be so scrupulous, John"*: Loretta McLaughlin, *The Pill, John Rock, and the Church: The Biography of a Revolution* (Boston: Little, Brown, 1982), p. 14.
104 *Rock's diary from 1907*: Margaret Marsh and Wanda Ronner, *The Fertility Doctor: John Rock and the Reproductive Revolution* (Baltimore, MD: Johns Hopkins University Press, 2008), p. 13.
105 *hustle between two exam rooms*: Rachel Achenbach, interview conducted by the author, October 2011.

105 *and he was "Dr. Rock"*: Transcript of Loretta McLaughlin interview conducted by Rachel Achenbach, undated, CLM.

105 *"a very poor scientist"*: Ibid.

106 *"without dire consequences"*: James Reed, *From Private Vice to Public Virtue: The Birth Control Movement and American Society Since 1830* (New York: Basic Books, 1978), p. 188.

106 *served as an ambulance driver*: McLaughlin, *The Pill, John Rock, and the Church*, p. 17.

107 *"natural fullness of ecstasy"*: John Rock, "Sex, Science and Survival," *Eugenics Review* 56, no. 2 (1964), p. 73.

107 *"shameful and intrinsically vicious"*: Janet E. Smith, *Humanae Vitae: A Generation Later* (Washington, DC: The Catholic University of America Press, 1991), p. 7.

109 *his controversial cause*: Leslie Woodcock Tentler, *Catholics and Contraception: An American History* (New York: Cornell University Press, 2004), p. 115.

109 *"ensure group survival"*: Ibid., pp. 77–78.

110 *"remain a Catholic"*: Margaret Sanger to Marion Ingersoll, February 18, 1954, Margaret Sanger Papers, SSC.

110 *"reformed Catholic"*: Katharine Dexter McCormick to Margaret Sanger, July 21, 1954, Margaret Sanger Papers, SSC.

111 *"I don't think that Roman Catholicism"*: "Planed Fertility," *Time*, February 9, 1948.

111 *"superstition, science, and symbolism"*: John Rock and David Loth, "Birth Control Is Not Enough," *Coronet*, June 1950, 67–72.

114 *reverse the procedure*: Marsh and Ronner, *The Fertility Doctor*, p. 131.

115 *"may be considered as deviants"*: Elaine Tyler May, *Barren in the Promised Land: Childless Americans and the Pursuit of Happiness* (Cambridge, MA: Harvard University Press, 1997), p. 172.

115 *"basic urge and need"*: Ibid., p. 153.

115 *"frustrated, but valiantly adventuresome"*: Marsh and Ronner, *The Fertility Doctor*, p. 154.

116 *chatted between sessions*: Albert Q. Maisel, *The Hormone Quest* (New York: Random House, 1965), p. 119.

116 *he wouldn't drop dead*: Marsh and Ronner, *The Fertility Doctor*, p. 155.

116 *careful not to make promises*: Transcript of Luigi Mastroianni interview conducted by Leon Speroff, undated.

116 *"they wanted to try it"*: McLaughlin, *The Pill, John Rock, and The Church*, p. 109.

117 *fifty milligrams of progesterone*: Laura V. Marks, *Sexual Chemistry* (New Haven, CT: Yale University Press, 2001), p. 93.

117 *"conception could not occur"*: Ibid., p. 110.

119 *nineteen shares valued at $921.50*: P. E. Tillman to Gregory Pincus, June 16, 1953, Gregory Pincus Papers, LOC.

119 *where it would solidify into a pellet*: Gregory Pincus to Al Raymond, November 16, 1953, Gregory Pincus Papers, LOC.

119 *gave him a green light*: Al Raymond to Gregory Pincus, November 12, 1953, Gregory Pincus Papers, LOC.

119 *not to publicize their involvement*: Gregory Pincus to Victor Drill, December 15, 1954, Gregory Pincus Papers, LOC.

119 *"going against Nature"*: McLaughlin, *The Pill, John Rock, and the Church*, p. 111.

CHAPTER ELEVEN

122 *"workings of the human body"*: Albert Q. Maisel, *The Hormone Quest* (New York: Random House, 1965), p. ix.

124 *"rising standard for the entire world"*: Andrea Tone, *Devices and Desires: A History of Contraceptives in America* (New York: Hill and Wang, 2001), p. 208.

124 *"tsunami of male lust"*: Mary Louise Roberts, *What Soldiers Do: Sex and the American GI in World War II France* (Chicago, IL: University of Chicago Press, 2013), p. 9.

125 *median age for marriage . . . was 20.1*: "American Families: 75 Years of Change," *Monthly Labor Review*, Bureau of Labor Statistics, March 1990, p. 7.

127 *"All animals play around"*: Beth Bailey, *Sex in the Heartland* (Cambridge, MA: Harvard University Press, 1999), p. 46.

CHAPTER TWELVE

128 *high hopes for a dramatic outcome*: Gregory Pincus, "Report of Progress," January 23, 1953, Gregory Pincus Papers, LOC.

129 *"patentable discoveries"*: Paul Henshaw to Gregory Pincus, January 26, 1953, Gregory Pincus Papers, LOC .

129 *"forward thinking required by research"*: Esther Katz, ed., *The Selected Papers of Margaret Sanger*, Vol. 3 (Urbana: University of Illinois Press, 2010), p. 349.

129 *"would agree to such a provision"*: Paul Henshaw to Gregory Pincus, January 26, 1953, Gregory Pincus Papers, LOC.

129 *"a knotty question"*: Gregory Pincus to Paul Henshaw, January, 28, 1953, Gregory Pincus Papers, LOC.

129 *can the human testing begin?*: Paul Henshaw to Gregory Pincus, February 17, 1953, Gregory Pincus Papers, LOC.

130 *"a little faster. . . . "*: Gregory Pincus to Paul Henshaw, February 19, 1953, Gregory Pincus Papers, LOC.

130 *"thirty to forty women"*: Gregory Pincus to Planned Parenthood, "Application for a Grant," April 29, 1953, Gregory Pincus Papers, LOC.

131 *"fundamental facts" . . . "available resources"*: Gregory Pincus to Paul Henshaw, March 30, 1953, Gregory Pincus Papers, LOC.

131 *"passed over the research laboratory"*: Gregory Pincus, *The Control of Fertility* (New York: Academic Press, 1965), p. 8.

133 *meet somewhere in the middle*: Transcript of Luigi Mastroianni interview conducted by Leon Speroff, undated.

133 *both of which were in Worcester*: "Dr. H. L. Kirkendall Dies in Worcester," *Lowell Sun*, May 9, 1955.

134 *250 and 300 milligrams a day*: Gregory Pincus to Henry Kirkendall, April 30, 1953, LOC.

134 *Worcester Foundation in Shrewsbury*: Dr. Henry Kirkendall, Jr., telephone interview conducted by the author, April 2013.

CHAPTER THIRTEEN

135 *"on our side or not"*: Russell Marker interview conducted by Jeffrey L. Sturchio, 1987 (Philadelphia Chemical Heritage Foundation, Oral History Transcript #0068).

137 *"a place I could work on them"*: Ibid.

138 *others were doing revolutionary work*: Carl Djerassi, *This Man's Pill: Reflections on the 50th Birthday of the Pill* (New York: Oxford University Press, 2001), p. 38.

138 *"funky little vacation house"*: Djerassi, *This Man's Pill*, p. 43.

139 *"Not in our wildest dreams"*: Andrea Tone, *Devices and Desires: A History of Contraceptives in America* (New York: Hill and Wang, 2001), p. 218.

CHAPTER FOURTEEN

140 *"the crowds are so great"*: Katharine Dexter McCormick to Margaret Sanger, May 15, 1953, Margaret Sanger Papers, SSC.

140 *hot and humid Monday*: Katharine Dexter McCormick to Margaret Sanger, Western Union telegram, June 1, 1953; "First Heat Wave Will End Today," *Lowell Sun*, June 8, 1953.

141 *"This is the place"*: Isabelle Chang, telephone interview conducted by the author, July 2013.

141 *half of the $17,500*: Paul Henshaw to Gregory Pincus, May 28, 1953, LOC.

141 *a check for $10,000*: Gregory Pincus to Paul Henshaw, June 10, 1953, LOC.

CHAPTER FIFTEEN

143 *because no work gets done*: Katharine Dexter McCormick to Margaret Sanger, December 27, 1954, Margaret Sanger Papers, SSC.

143 *"the scope of the tests in action"*: Katharine Dexter McCormick to Margaret Sanger, September 28, 1953, Margaret Sanger Papers, SSC.

143 *to fund Pincus's research beyond January 1954*: Margaret Sanger to Katharine Dexter McCormick, October 5, 1953, Armond Fields Collection, USC.

143 *power struggle with William Vogt*: Margaret Sanger to Marion Crary Ingersoll, February 18, 1954, Margaret Sanger Papers, SSC.

143 *"a simple, cheap, contraceptive"*: Margaret Sanger to Katharine Dexter McCormick, February 23, 1954, Margaret Sanger Papers, SSC.

144 *"development of a simple contraceptive"*: Ibid.

144 *"bring it to a final conclusion"*: Margaret Sanger to Katharine Dexter McCormick, October 12, 1953, Armond Fields Collection, USC.

144 *"a bit of luck these days"*: Gregory Pincus to Al Raymond, May 8, 1953, Gregory Pincus Papers, LOC.

145 *Syntex had performed better*: Transcript of Gabriel Bialy interview conducted by Leon Speroff, August 2007.

145 *8 percent of the Foundation's total income*: "Tenth Anniversary Report, 1944–1954," Worcester Foundation for Experimental Biology, LOC.

145 *one-third of Pincus's $15,000 annual salary*: Worcester Foundation for Experimental Biology Finance Committee Report, November 3, 1953, Worcester Foundation Papers, UM.

145 *"weary & depressed"*: Margaret Sanger to Juliet Barrett Rublee, January 26, 1953, Margaret Sanger Papers, SSC.

146 *she wrote to the same friend*: Ibid.

146 *"not to do any public speaking ever again"*: Margaret Sanger to Dorothy Hamilton Brush, January 14, 1952, Margaret Sanger Papers, SSC.

146 *urged her to retire*: Esther Katz, ed., *The Selected Papers of Margaret Sanger*, Vol. 3 (Urbana: University of Illinois Press, 2010), p. 345.

146 *"Preposterous!"*: Margaret Sanger to Rufus Day, Jr., December 6, 1956, Margaret Sanger Papers, SSC.

146 *to which she had become accustomed*: Katz, *The Selected Papers of Margaret Sanger*, Vol. 3, p. 319.

146 *Cleveland-born socialite*: Dorothy Hamilton Brush to Margaret Sanger, January 6, 1953, Margaret Sanger Papers, SSC.

147 *"when the slavery of half of humanity"*: Simone de Beauvoir, *The Second Sex* (New York: Random House, 2011), p. 766.

147 *"a life time to study & write"*: Margaret Sanger to Juliet Barrett Rublee, February 6, 1953, Margaret Sanger Papers, SSC.

147 *"put all our energies into research"*: *Fourth International Conference on Planned Parenthood, Report of the Proceedings*, August 17–22, 1953, Stockholm, Sweden (London: International Planned Parenthood Federation, 1953), p. 9.

148 *when individuals volunteered for sterilization*: Irene Headley Armes, "A

Proposed Program of Research on the Status and Social Demand for Permanent Conception Control in the U.S.A," *Fourth International Conference on Planned Parenthood*, August 17–22, 1953, Stockholm, Sweden.

148 *"both the individual and the community"*: Ibid.

149 *"More children from the fit"*: "Intelligent or Unintelligent Birth Contol," *Birth Control Review*, May 1919, p. 12.

149 *just as immigrants applied for visas*: Speech by Margaret Sanger in Hartford, Connecticut, February 11, 1923, Margaret Sanger Papers, SSC.

149 *"were the government not feeding them"*: Margaret Sanger to Katharine Dexter McCormick, October 27, 1950, Armond Fields Collection, USC.

149 *"a privilege, not a right"*: Katz, *The Selected Papers of Margaret Sanger*, Vol. 3, p. 271.

150 *"conservative program of social control"*: David M. Kennedy, *Birth Control in America* (New Haven, CT: Yale University Press, 1970), p. 121.

151 *nothing physical between them:* Lawrence Lader, "Margaret Sanger: Militant Pragmatist Visionary," *On The Issues*, Spring 1990, http://www.ontheissuesmagazine.com/1990spring/Spr90_Lader.php (accessed February 18, 2014).

151 *"I am not happy in past memories"*: Katz, *The Selected Papers of Margaret Sanger*, Vol. 3, p. 333.

151 *"inexhaustible flame of your own driving force?"*: Lawrence Lader to Margaret Sanger, July 25, 1952, Margaret Sanger Papers, SSC; Katz, *The Selected Papers of Margaret Sanger*, Vol. 3, p. 333.

152 *"could make you quite ill"*: Ibid., p. 344.

152 *"its prophet, its driving force"*: Lawrence Lader, *The Margaret Sanger Story* (New York: Doubleday, 1955), p. 340.

CHAPTER SIXTEEN

153 *a total of $622,000 in income*: Bruce Crawford, Worcester Foundation for Experimental Biology Finance Committee Report, October, 16, 1953, Worcester Foundation Papers, UM.

153 *"to do the jobs on hand"*: Ibid.

154 *pledging fifty thousand dollars*: Ibid.

154 *"young, lusty and full of promise"*: Gregory Pincus to Frank Fremont-Smith, September 25, 1953, LOC.

155 *metabolism of steroids*: "Synopsis of Worcester Foundation for Experimental Biology Research Projects," 1953–54, Worcester Foundation Papers, UM.

155 *"new studies in reproduction control"*: "Minutes of the Tenth Annual Trustees Meeting of the Worcester Foundation for Experimental Biology," June 12, 1954, Worcester Foundation Papers, UM.

155 *in a letter to McCormick*: Margaret Sanger to Katharine Dexter McCormick, February 13, 1954, Margaret Sanger Papers, SSC.

155 *"merely said he hoped I was still interested"*: Katharine Dexter McCormick to Margaret Sanger, February 17, 1954, Armond Fields Collection, USC.
155 *"As I became somewhat impatient"*: Ibid.
156 *"I was mistaken"*: Ibid.
156 *first human trials*: Katharine Dexter McCormick to Margaret Sanger, November 13, 1953, Armond Fields Collection, USC.
157 *"forget to take the medicine sometimes"*: Ibid.
158 *"would be very difficult in this country"*: Gregory Pincus to Katharine Dexter McCormick, March 5, 1954, Gregory Pincus Papers, LOC.
158 *"somewhat elevated"*: Gregory Pincus to Al Raymond, January 26, 1954, Gregory Pincus Papers, LOC.
158 *desperate for better birth control*: Annette B. Ramirez de Arellano, *Colonialism, Catholicism, and Contraception* (Chapel Hill: University of North Carolina Press, 1983), p. 108.
159 *"ovulating intelligent" women*: Katharine Dexter McCormick to Margaret Sanger, October 21, 1954, Margaret Sanger Papers, SSC.
160 *average mother . . . had borne 6.8 children*: Reuben Hill, J. Mayone Stycos, and Kurt W. Black, *The Family and Population Control* (Chapel Hill: University of North Carolina Press, 1959), p. 13.
161 *"determined not to have more children"*: Transcript of Edris Rice-Wray interview conducted by Ellen Chesler, undated, SSC.

CHAPTER SEVENTEEN

162 *ten times as many children*: David M. Oshinsky, *Polio: An American Story* (Oxford, UK, and New York: Oxford University Press, 2005), p. 5.
163 *six hundred thousand children*: Ibid., p. 199.
163 *"The earth's population will double"*: "There Won't be Room to Breathe in 2023 If Birth-Death Rate Continues," *Panama City News-Herald*, April 6, 1953.
164 *one of the most densely populated countries*: "Population Control in Puerto Rico: The Formal and Informal Framework," *Law and Contemporary Problems* 25, no. 3 (1960), pp. 558–76.
164 *more densely packed than the United States*: Ibid.
164 *one in every ten residents*: "Flow of Puerto Ricans Here Fills Jobs, Poses Problems," *New York Times*, February 23, 1953.
164 *one hundred miles long*: "Puerto Rico Faces Two Big Problems," *New York Times*, June 27, 1954.
165 *only heightening the sense of crowding*: Ibid.
165 *less than four*: P. K. Hatt, *Backgrounds of Human Fertility in Puerto Rico* (Princeton, NJ: Princeton University Press, 1952), p. 53, Table 37.
165 *"So, two is enough"*: J. Mayone Stycos, *Family and Fertility in Puerto Rico* (New York: Columbia University Press, 1955), p. 160.

165 *underway in Puerto Rico*: Ibid., p. 159.
166 *"How could I enjoy it . . . ?"*: Ibid., pp. 163–64.
166 *women . . . intentionally married men*: Ibid., p. 164.
167 *birth control at some point*: Ibid., p. 217.
167 *a return flight home on Monday*: Annette B. Ramirez de Arellano, *Colonialism, Catholicism, and Contraception* (Chapel Hill: University of North Carolina Press, 1983), p. 146.
167 *would cost about six hundred dollars*: Ibid.
168 *"imperils the whole society"*: Stycos, *Family and Fertility in Puerto Rico*, p. 255.
168 *" 'but you won't do both' "*: Transcript of Edris Rice-Wray interview conducted by Ellen Chesler, undated, Margaret Sanger Papers, SSC.
169 *They recommended a diaphragm*: Transcript of Edris Rice-Wray oral history by James Reed, March 1987, Margaret Sanger Papers, SSC.
169 *came as a revelation*: Transcript of Edris Rice-Wray interview conducted by Ellen Chesler, undated, Margaret Sanger Papers, SSC.
169 *"it wasn't enough for me"*: Ibid.
169 *to help the other women*: Ibid.
169 *she moved with her children to San Juan*: Ibid.
170 *"they are doing nothing about it"*: Edris Rice-Wray to William Vogt, December 10, 1953, Gregory Pincus Papers, LOC.
170 *"We have 160 clinics"*: Ibid.
171 *"looking for anybody"*: Transcript of Edris Rice-Wray interview conducted by Ellen Chesler, undated, Margaret Sanger Papers, SSC.
171 *"Our great opportunity"*: Edris Rice-Wray to Gregory Pincus, March 6, 1954, Gregory Pincus Papers, LOC.
172 *Only five reported no side effects*: Memo titled "Pseudopregnancy Data," June 15, 1954, Gregory Pincus Papers, LOC.
172 *"assemble a group of 50 women"*: Gregory Pincus to Dr. Manuel Fernández Fuster, October 19, 1954, LOC.
172 *"at worst inconvenient"*: Gregory Pincus memo, November 1, 1954, John Rock Papers, CLM.

CHAPTER EIGHTEEN

175 *Mercier said*: Tina Mercier, telephone interview conducted by the author, April 2013.
176 *behind his back*: Dr. Enoch Callaway, telephone interview conducted by the author, March 2013.
177 *one inmate beheaded another*: "12-State Alarm for Worcester Mental Patient Who Axed Inmate," *Lowell Sun*, July 22, 1943.
177 *"patients who have defeated our best efforts"*: Enoch Callaway, *Asylum:*

A Mid-Century Madhouse and Its Lessons about Our Mentally Ill Today (Westport, CT: Praeger, 2007), p. 6.

178 *"defecating and urinating"*: Ibid., p. 9.

178 *pulling hair in frustration*: Ibid., p. 8.

178 *"I could not help imagining"*: Ibid., p. 9.

178 *"would never think of doing these days"*: Dr. Enoch Callaway, telephone interview conducted by the author, March 2013.

179 *"[W]e wish to inform the directors"*: Oscar Resnick to Gregory Pincus, undated, Gregory Pincus Papers, LOC.

180 *to cure men of homosexuality*: Andrea Tone, ed., *Controlling Reproduction* (Wilmington, DE: Scholarly Resources, 2001), p. 220.

180 *"were just as psychotic"*: "Field Study with Enovid as a Contraceptive Agent," ERW, *Proceedings of a Symposium on 19-Nor Progestational Steroids*, 118, Searle Research Laboratories, January 23, 1957.

CHAPTER NINETEEN

181 *it made him uncomfortable*: Rachel Achenbach, interview conducted by the author, October 2011.

181 *hormone for prolonged stretches*: Loretta McLaughlin, *The Pill, John Rock, and The Church: The Biography of a Revolution* (Boston: Little, Brown, 1982), p. 111.

182 *"very realistic about medical work"*: Katharine Dexter McCormick to Margaret Sanger, July 19, 1954, Margaret Sanger Papers, SSC.

182 *Pincus and Rock refused*: Margaret Marsh and Wanda Ronner, *The Fertility Doctor: John Rock and the Reproductive Revolution* (Baltimore, MD: Johns Hopkins University Press, 2008), p. 158.

182 *"an abstract research thing"*: Ibid., p. 159.

182 *"What has happened to you . . . ?"*: Margaret Sanger to Abraham Stone, March 2, 1954, Margaret Sanger Papers, SSC.

184 *"pitifully little that was of practical value"*: Ibid., p. 169.

184 *keynote speaker was Catholic?*: Winfield Best to John Rock, March 9, 1954, John Rock Papers, CLM.

184 *"importance of the world population increase"*: John Rock to Winfield Best, March 11, 1954, John Rock Papers, CLM.

186 *"Two big steps that women must take"*: David Halberstam, *The Fifties* (New York: Villard Books, 1993), p. 591.

186 *wash the dishes and emerge "utterly desirable"*: Marlene Dietrich, "How to Be Loved," *Ladies' Home Journal*, January 1954.

186 *"learn to live and work together"*: "Listen, Marlene!" *Ladies' Home Journal*, April 1954.

187 *behind the scenes on political campaigns*: Joanne Meyerowitz, ed., *Not*

June Cleaver: Women and Gender in Postwar America, 1945–1960 (Philadelphia, PA: Temple University Press, 1994), p. 250.

187 *"end of the old controversy"*: James R. Petersen, *The Century of Sex: Playboy's History of the Sexual Revolution, 1900–1999* (New York: Grove Press, 1999), p. 233.

188 *"Sex is something I really don't understand"*: J. D. Salinger, *The Catcher in the Rye* (Boston: Little, Brown, 1951), pp. 63–64.

188 *"wearing out britches from the inside"*: Ibid., p. 240.

188 *having an affair with a nurse*: Gay Talese, *Thy Neighbor's Wife* (New York: Doubleday, 1980), p. 50.

188 *phone call from the chancellery*: Ibid., p. 72.

189 Playboy *was the fastest growing magazine in America*: Ibid., p. 73.

189 *"develop a method for inhibition"*: Gregory Pincus to Margaret Sanger, March 31, 1954, Gregory Pincus Papers, LOC.

CHAPTER TWENTY

190 *shaken but not hurt*: Katharine Dexter McCormick to Margaret Sanger, February 1, 1955, Margaret Sanger Papers, SSC.

190 *hitched a ride to Boston*: Ibid.

190 *maid stood by to fetch drinks*: Laura Pincus Bernard, interview conducted by the author, October 2011.

191 *refused to supply the chemical*: Margaret Sanger to Katharine Dexter McCormick, April 22, 1954, Margaret Sanger Papers, SSC.

191 *did not yet understand how or why*: Margaret Marsh and Wanda Ronner, *The Fertility Doctor: John Rock and the Reproductive Revolution* (Baltimore, MD: Johns Hopkins University Press, 2008), p. 170.

191 *"throws grave doubt"*: Al Raymond to Gregory Pincus, January 3, 1955, Gregory Pincus Papers, LOC.

192 *"will send it to you unlabeled"*: Ibid.

192 *"didn't want to be bothered with menstruals"*: Transcript of Anne Merrill interview conducted by Leon Speroff, May 2007.

193 *participate as part of their studies*: Katharine Dexter McCormick to Margaret Sanger, February 1, 1955, Margaret Sanger Papers, SSC.

194 *Laura was startled and charmed*: Laura Pincus Bernard, interview conducted by the author, October 2011; David Halberstam, *The Fifties* (New York: Villard Books, 1993), p. 604.

194 *seeing patients and conducting experiments*: Armond Fields, *Katharine Dexter McCormick: Pioneer for Women's Rights* (Westport, CT: Praeger, 2003), p. 268.

194 *confident enough to send Pincus a check*: Katharine Dexter McCormick to Bruce Crawford, January 5, 1955, Worcester Foundation Papers, UM.

194 *in addition to the $20,000*: Katharine Dexter McCormick to Hudson Hoagland, August 13, 1954, Worcester Foundation Papers, UM.
195 *"for lack of funds"*: Katharine Dexter McCormick to Bruce Crawford, January 5, 1955, Worcester Foundation Papers, UM.
195 *food supplies might falter*: "60,000,000 Buyers to Enter Market," *New York Times*, March 15, 1955.
196 *"Babies, Babies, Babies—4,000,000 Problems"*: "Washington: Babies, Babies, Babies—4,000,000 Problems," *New York Times*, February 27, 1955.
197 *"meeting the needs of the people"*: "San Juan Talks Open on Birth Control Theme Held Key to Caribbean Problems," *New York Times*, May 13, 1955.
198 *"I wouldn't have your job for anything"*: "The Plight of the Young Mother," *Ladies' Home Journal*, February 1956, p. 107.
198 *"if you call that a vacation"*: Ibid.
199 *"very quietly and privately"*: "Scientists Near Goal in Finding Simple Birth Control Method," *Middleboro Daily News* (Kentucky), July 14, 1955.
199 *project-by-project basis*: Ellen Chesler, *Woman of Valor: Margaret Sanger and the Birth Control Movement in America* (New York: Simon & Schuster, 2007), p. 437.
199 *Conference of the International Planned Parenthood Federation*: Ibid.
200 *"I do wish the field tests"*: Katharine Dexter McCormick to Margaret Sanger, February 1, 1955, Margaret Sanger Papers, SSC.

CHAPTER TWENTY-ONE

201 *did not bother him*: Laura Pincus Bernard, interview conducted by the author, July 2013.
201 *"hold it against her"*: David Tyler to Gregory Pincus, July 8, 1955, Gregory Pincus Papers, LOC.
202 *"it will not succeed"*: David Tyler to Gregory Pincus, June 14, 1955, Gregory Pincus Papers, LOC.
203 *which she took for insomnia*: Jean H. Baker, *Margaret Sanger: A Life of Passion* (New York: Hill and Wang, 2011), p. 285.
203 *"now I do not need anything"*: Margaret Sanger to Juliet Barrett Rublee, February 13, 1955, Margaret Sanger Papers, SSC.
203 *"veins of sadness"*: Ellen Chesler, *Woman of Valor: Margaret Sanger and the Birth Control Movement in America* (New York: Simon & Schuster, 2007), p. 415.
204 *638,000 legal abortions*: "Mrs. Sanger's Visit Excites Japanese," *New York Times*, November 10, 1952.
204 *demand for abortions would decline*: "Foreword," *Fifth International Conference on Planned Parenthood, Report of the Proceedings* (Tokyo), October 1955, Margaret Sanger Papers, SSC.

204 *"more scientific" titles*: Margaret Sanger to Katharine Dexter McCormick, April 13, 1955, Margaret Sanger Papers, SSC.

204 *"be progesterone guinea pigs"*: Ibid.

204 *"evidently a very necessary help to him"*: Ibid.

205 *"It looks pretty good"*: Margaret Marsh and Wanda Ronner, *The Fertility Doctor: John Rock and the Reproductive Revolution* (Baltimore, MD: Johns Hopkins University Press, 2008), p. 170.

206 *"a fair drop off"*: Gregory Pincus to Katharine Dexter McCormick, October 1, 1955, Margaret Sanger Papers, SSC.

206 *between San Juan and Shrewsbury*: Gregory Pincus to Celso Garcia, June 23, 1955, Margaret Sanger Papers, SSC.

206 *"very little [data] worth reporting"*: Gregory Pincus to David Tyler, June 22, 1955, Margaret Sanger Papers, SSC.

207 *charged her purchases to McCormick*: Assorted receipts, Gregory Pincus Papers, LOC.

207 *furniture from an uncle in Montreal*: Assorted receipts, Worcester Foundation Papers, UM.

208 *his upcoming travels*: Katharine Dexter McCormick to Bruce Crawford, July 12, 1955, Worcester Foundation Papers, UM.

208 *stop work on the pill*: Katharine Dexter McCormick to Margaret Sanger, June 29, 1955, Margaret Sanger Papers, SSC.

208 *"You talk to young college women"*: "Margaret Sanger Thinks Crusading Spark Dampened," *Oxnard Press Courier*, May 10, 1955.

208 *conditions of women in prison*: Ibid.

209 *attended the conference in Japan*: Beryl Suitters, *Be Brave and Angry: Chronicles of the International Planned Parenthood Federation* (London: International Planned Parenthood Federation, 1973), p. 132.

CHAPTER TWENTY-TWO

210 *two thousand U.S. civilians*: National World War II Museum, http://www.nationalww2museum.org/learn/education/for-students/ww2-history/ww2-by-the-numbers/world-wide-deaths.html (accessed February 18, 2014).

211 *"last immodest exercise"*: John Dower, *Embracing Defeat: Japan in the Wake of World War II* (New York: W. W. Norton & Company, 1999), p. 23.

211 *courtesans, prostitutes, military pawns*: Michael Hoffman, "Revolution Was in the Air During Japan's Taisho Era," *Japan Times*, July 29, 2012, http://www.japantimes.co.jp/life/2012/07/29/general/revolution-was-in-the-air-during-japans-taisho-era-but-soon-evaporated-into-the-status-quo/#.UwdeDpGuPk4 (accessed February 20, 2014).

211 *thirteen-hour shifts*: Sanger Diary, 1922, Margaret Sanger Papers, SSC.

212 *"vivid and long-enduring impression"*: Ellen Chesler, *Woman of Valor: Margaret Sanger and the Birth Control Movement in America* (New York: Simon & Schuster, 2007), p. 365.

212 *illegal abortion-inducing medicine*: Carolyn Eberts, "The Sanger Brand: The Relationship of Margaret Sanger and the Pre-War Japanese Birth Control Movement," Master's Thesis, Bowling Green State University, 2010.

212 *"no priests denouncing me"*: Margaret Sanger, *My Fight for Birth Control* (New York: Farrar & Rinehart Incorporated, 1931), p. 254.

213 *abortion rates in the country rose sharply*: Sheila Matsumoto, "Women in Factories" in *Women in Changing Japan*, ed. Joyce Lebra, Joy Paulson, and Elizabeth Powers (Boulder, CO: Westview Press, 1970), p. 56.

213 *would jump from 6,000*: Yoshio Koya, *Pioneering in Family Planning* (Tokyo: Japan Medical Publishers, 1963), p. 63.

214 *"will eliminate contraceptive devices"*: "Birth-control Pill Reported by Expert," *Pasadena Independent*, October 19, 1955.

214 *"the miracle tablet maybe"*: "Foreword," *Fifth International Conference on Planned Parenthood, Report of the Proceedings* (Tokyo), October 1955, SSC.

214 *he complained of stomach trouble*: Laura Pincus Bernard's scrapbook, Pincus family collection.

215 *she felt as if she were at home*: Headline unavailable, *Mainichi Shimbun*, October 24, 1955.

215 *"talk about in the previous conferences"*: Headline unavailable, *Asahi Shimbun*, October 24, 1955.

216 *"no such substance is yet known"*: Paul Vaughan, *The Pill on Trial* (New York: Coward-McCann, Inc., 1970), pp. 32–33.

216 *time to prove it really worked*: Transcript of Judy McCann interview conducted by Leon Speroff, May 2007.

216 *were a "necessary evil"*: Vaughan, *The Pill on Trial*, p. 42.

216 *"Unless and until we know more"*: Ibid., p. 33.

217 *"He was the most supremely confident"*: Laura Pincus Bernard, e-mail to the author, September 1, 2013.

217 *"We cannot on the basis of our observations"*: Vaughan, *The Pill on Trial*, p. 33.

218 *" as close as we should like"*: Ibid., p. 34.

218 *"We need better evidence"*: Ibid.

CHAPTER TWENTY-THREE

219 *"the magic and mystery of our time"*: Gregory Pincus to Hermann Joseph Muller, Hermann Joseph Muller Papers, Lilly Library, Indiana University, Bloomington, Indiana.

220 *voting in roughly equal numbers*: "Women's Vote: The Bigger Half?" *New York Times Magazine*, October 21, 1956.

221 *"Just Darn Mad"*: "Letters to Geraldine," *Oakland Tribune*, November 3, 1955.

221 *"poor choice as a reward"*: Ibid.

221 *"Someone ought to inform this young lady"*: "Letters to Geraldine," *Oakland Tribune*, December 8, 1955.

221 *"Can anyone say I'm a sinner"*: Ibid.

222 *"I'd've fucked anything"*: David Dalton, *Piece of My Heart: The Life, Times and Legend of Janis Joplin* (New York: St. Martin's Press, 1985), p. 147.

222 *"rat race or domesticity"*: Marge Piercy, "Through the Cracks: Growing Up in the Fifties," in *Particolored Blocks for a Quilt* (Ann Arbor: University of Michigan Press, 1982), pp. 155–56.

222 *"good and hard"*: Grace Metalious, *Peyton Place* (New York: Julian Messner, 1956), p. 124.

223 *one in twenty-nine Americans*: Ibid., p. viii.

223 *"something going on out there"*: Ibid., p. ix.

223 *"declared out-of-bounds"*: Ibid., p. xiv.

224 *to promote sex education*: Linda Gordon, *The Moral Property of Women: A History of Birth Control Politics in America* (Urbana: University of Illinois Press, 2007), p. 255.

224 *"who want to see a change"*: "The Attitude of the Roman Catholic Church," Internal Memo, International Planned Parenthood Federation, February 28, 1955, Margaret Sanger Papers, SSC.

225 *"law which is natural and divine"*: John T. Noonan, *Contraception: A History of Its Treatment by the Catholic Theologians and Canonists* (Cambridge, MA: Harvard University Press), p. 467.

226 *average was about 20 percent higher*: Leslie Woodcock Tentler, *Catholics and Contraception: An American History* (Ithaca, NY: Cornell University Press, 2004), p. 133.

226 *"a comfortable future on earth"*: Ibid., p. 132.

226 *decided not to publish the results*: Ibid., p. 200.

226 *sacraments of confession and communion*: Ibid., p. 135.

226 *"despite frantic and distressing efforts"*: Anonymous letter to the editor, *Liguorian* 48, no. 10 (1960), p. 39.

227 *"I was taught by the Church"*: Loretta McLaughlin, interview conducted by the author, October 2011.

227 *"high-handedness"*: Ibid.

227 *to find out if they worked*: Katharine Dexter McCormick, "Notes on Conversation with John Rock," January 9, 1956, Margaret Sanger Papers, SSC.

228 *"'don't you sell my church short'"*: Loretta McLaughlin, *The Pill, John Rock, and the Church* (Boston: Little, Brown, 1982), p. 142.

CHAPTER TWENTY-FOUR

229 *He needed a new approach*: Katharine Dexter McCormick, "Notes on Conversation with Dr. Pincus," March 5, 1956, Armond Fields Collection, USC.

230 *"I was kind of scared"*: Transcript of Edris Rice-Wray interview conducted by Ellen Chesler, undated.

230 *so charming and self-assured*: Ibid.

230 *"none of the Church's damn business"*: Ibid.

231 *stable over months of work*: Edris Rice-Wray to Gregory Pincus, April 17, 1956, Gregory Pincus Papers, LOC.

231 *manage their family's size*: Edris Rice-Wray, "Field Study with Enovid as a Contraceptive Agent," *Proceedings of a Symposium on 19-Nor Progestational Steroids*, 79, Searle Research Laboratories, January 23, 1957.

231 *"very easy to talk to mothers"*: Edris Rice-Wray, Speech to the Royal Swedish Endocrine Society, March 9, 1962, Gregory Pincus Papers, LOC.

232 *religion prohibited her from enrolling*: Ibid.

232 *"crazy to get the pill"*: Edris Rice-Wray to Gregory Pincus, April 17, 1956, Gregory Pincus Papers, LOC.

233 *"We are led to suspect"*: Loretta McLaughlin, *The Pill, John Rock, and the Church* (Boston: Little, Brown, 1982), p. 122.

233 *he recalled years later*: Ibid., p. 123.

234 *"my results were all wrong"*: "Recent Progress in Hormone Research," *Proceedings of the Laurentian Hormone Conference*, Vol. 13 (New York: Academic Press Inc., 1957), p. 340.

234 *"superior forms of entertainment"*: McLaughlin, *The Pill, John Rock, and the Church*, p. 45.

234 *plunged nude into the swimming pool*: Ibid.

CHAPTER TWENTY-FIVE

235 *"things began to happen"*: Margaret Sanger to Katharine Dexter McCormick, December 12, 1956, Margaret Sanger Papers, SSC.

235 *"the conspiracy of silence is broken"*: Ibid.

236 *daiquiris in bed in the morning*: Madeline Gray, *Margaret Sanger: A Biography of the Champion of Birth Control* (New York: Richard Marek Publishers, 1979), p. 429.

236 *put her to bed*: Ibid., p. 428.

236 *"woman's biological freedom and development"*: Margaret Sanger to Dr. Kenneth Rose, August 20, 1956, Margaret Sanger Papers, SSC.

236 *"that inane one Planned Parenthood"*: Ibid.

237 *it was going to be big*: Geoff Dutton, interview conducted by the author, October 2011.

237 *"until I hear from you"*: Peggy Blake to Gregory Pincus, July 28, 1956, Gregory Pincus Papers, LOC.

237 *"pretty much persuades me"*: Gregory Pincus to Peggy Blake, August 2, 1956, Gregory Pincus Papers, LOC.

238 *"I was ready to murder"*: Peggy Blake to Gregory Pincus, August 4, 1956, Gregory Pincus Papers, LOC.

238 *"lose money on the deal"*: Ibid.

238 *"start again" on a new bottle*: Edris Rice-Wray, "Field Study with Enovid as a Contraceptive Agent," *Proceedings of a Symposium on 19-Nor Progestational Steroids*, 79, Searle Research Laboratories, January 23, 1957.

238 *took all the pills at once*: Laura Pincus Bernard, interview conducted by the author, July 2013.

239 *"A woman dressed as a nurse"*: Edris Rice-Wray's translation of article from *El Imparcial*, April 21, 1956, Gregory Pincus Papers, LOC.

239 *experiencing unpleasant side effects*: Iris Rodriguez to Gregory Pincus, May 8, 1956, Gregory Pincus Papers, LOC.

239 *twenty of the original one hundred*: Gregory Pincus to Katharine Dexter McCormick, October 11, 1956, Gregory Pincus Papers, LOC.

239 *"'they are afraid of you'"*: Edris Rice-Wray to Gregory Pincus, December 20, 1956, Gregory Pincus Papers, LOC.

240 *she would have to drop it*: Ibid.

240 *"We will only say"*: Iris Rodriguez to Gregory Pincus, May 8, 1956, Gregory Pincus Papers, LOC.

240 *how they could get it*: Transcript of Edris Rice-Wray interview conducted by Ellen Chesler, undated, SSC.

240 *"calling on me when I make the visits"*: Ibid.

240 *"a major convincing influence for others"*: Celso-Ramón Garcia, M.D., "The Early History of Oral Contraceptives," draft of paper to be presented at the John Rock Commemorative Symposium, October 21, 1980, CLM.

241 *side effects were becoming too much*: Gregory Pincus to Katharine Dexter McCormick, October 11, 1956, Gregory Pincus Papers, LOC.

241 *seventeen days after delivering*: Edris Rice-Wray, "Field Study with Enovid as a Contraceptive Agent."

CHAPTER TWENTY-SIX

244 *"Sex before marriage?"*: Sue Dixon, interview conducted by the author, June 2013.

244 *"I took her up to Michigan"*: Wes Dixon, interview conducted by the author, June 2013.
246 *at heart he was a risk taker*: Sue Dixon, interview conducted by the author, June 2013.
246 *a higher percentage of its revenue*: William I. Latourette, "More Wonder Drugs," *Barron's National Business and Financial Weekly*, April 28, 1958, p. 11.
247 *on sales of $26 million in 1955*: "Container Corp. Sets 2 Records," *New York Times*, February 3, 1956, p. 31.
247 *Sue Dixon's husband, Wes*: "Searle Aide Promoted to Overseas Unit Post," *New York Times*, January 12, 1956.
247 *"Me?"*: Sue Dixon, interview conducted by the author, June 2013.
248 *grasped the pill's "sociological implications"*: Celso-Ramón Garcia, M.D., "The Early History of Oral Contraceptives," draft of paper to be presented at the John Rock Commemorative Symposium, October 21, 1980, CLM.
248 *the cause of population control*: Loretta McLaughlin, *The Pill, John Rock, and the Church* (Boston: Little, Brown, 1982), p. 135.
248 *omit the company's name*: Al Raymond to Gregory Pincus, October 4, 1957, Gregory Pincus Papers, LOC.

CHAPTER TWENTY-SEVEN

249 *"emotional super-activity of Puerto Rican women"*: Annette B. Ramirez de Arellano, *Colonialism, Catholicism, and Contraception* (Chapel Hill: University of North Carolina Press, 1983), p. 116.
250 *train a doctor in how to use it*: Ibid., p. 117.
251 *"O, doctora, opéreme"*: Transcript of Adaline Satterthwaite interview conducted by James Reed, June 1974, Schlesinger-Rockefeller Oral History Project.
251 *she would not sterilize*: Ibid.
251 *scarcely room "for a squeezed pedestrian"*: Clarence Gamble to Margaret Sanger, March 13, 1957, Margaret Sanger Papers, SSC.
251 *birth control they were using, if any*: Transcript of Adaline Satterthwaite interview conducted by James Reed, June 1974, Schlesinger-Rockefeller Oral History Project.
251 *"the cervix looks angry"*: Adaline Satterthwaite to Clarence Gamble, December 2, 1959, Clarence Gamble Papers, CLM.
253 *"In the 1,279 cycles"*: Nelly Oudshoorn, *Beyond the Natural Body* (London and New York: Routledge, 1994), p. 132.
254 *the babies that were arriving*: Albert Q. Maisel, *The Hormone Quest* (New York: Random House, 1965), p. 46.

255 *fetch her diaphragm from the bathroom*: Loretta McLaughlin, *The Pill, John Rock, and the Church* (Boston: Little, Brown, 1982), p. 138.

256 *taking control of the supply chain*: Gregory Pincus to Jack Searle, January 29, 1957, Gregory Pincus Papers, LOC.

256 *some of its new hormone products*: Remarks by J. G. Searle at Annual Meeting of Stockholders, April 26, 1958, transcript in Worcester Foundation records, UM.

256 *Djerassi urged Parke-Davis to fight*: Carl Djerassi, *This Man's Pill* (Oxford, UK: Oxford University Press, 2001), p. 54.

256 *"small potatoes"*: Ibid.

256 *"absolutely unexplored ground"*: Paul Vaughan, *The Pill on Trial* (New York: Coward-McCann, Inc., 1970), p. 49.

257 *"field of physiological birth control"*: "Contraceptive Pill?" *Time* magazine, May 5, 1957, p. 83.

CHAPTER TWENTY-EIGHT

258 *"Believed to have magical powers"*: Suzanne White Junod and Lara Marks, "Women's Trials: The Approval of the First Oral Contraceptive in the United States and Great Britain," *Journal of the History of Medicine and Allied Sciences* 57, no. 2 (2002), p. 127.

259 *"SUGGEST BUTTONING UP," it read*: Gregory Pincus to John Rock, June 26, 1957, Gregory Pincus Papers, LOC.

259 *"brilliant and painstaking work"*: "New Hope for Childless Women," *Ladies' Home Journal*, August 1957, p. 46.

259 *nor cooperated with the author*: I. C. Winter to Edward Tyler, August 6, 1957, Gregory Pincus Papers, LOC.

260 *only to doctors on the West Coast*: Gregory Pincus to Margaret Sanger, July 22, 1957, Gregory Pincus Papers, LOC.

260 *"any physician may prescribe it"*: Ibid.

260 *"While we are convinced"*: Ibid.

261 *give himself a treat*: Laura Pincus Bernard and Geoff Dutton, interviews with the author, October 2011.

262 *"safe for this purpose in short term medication"*: White Junod and Marks, "Women's Trials," pp. 117–60.

264 *"I beg you please help me if you can"*: Letter to Gregory Pincus, October 31, 1957, Gregory Pincus Papers, LOC.

264 *"To save my married daughter"*: Letter to Gregory Pincus, June 21, 1957, Gregory Pincus Papers, LOC.

264 *"I absolutely need your help"*: Letter to Gregory Pincus, undated, Gregory Pincus Papers, LOC.

265 *"a powerful weapon"*: Hugh Hefner, telephone interview conducted by the author, February 2012.

265 *"I don't remember that"*: Ibid.
265 *"It was like a free ad"*: McLaughlin, *The Pill, John Rock, and the Church*, p. 139.

CHAPTER TWENTY-NINE

266 *"Good evening"*: Transcript of Margaret Sanger interview conducted by Mike Wallace in Esther Katz, ed., *The Selected Papers of Margaret Sanger,* Vol. 3 (Urbana: University of Illinois Press, 2010), p. 423.
266 *unusually warm and humid day*: "Weather Man Slips Up on Ice Skating in City," *New York Times*, September 22, 1957.
266 *"nosy, irreverent, often confrontational"*: Mike Wallace, *Between You and Me* (New York: Hyperion, 2005), p. 2.
266 *"an intrepid gadfly who had the temerity"*: Ibid., p. 136.
267 *"May I ask you this?"*: Transcript of Margaret Sanger interview conducted by Mike Wallace in Katz, ed., *Selected Papers of Margaret Sanger*, Vol. 3, p. 432.
268 *"It is just too bad"*: Letter to Margaret Sanger, September 21, 1957, Margaret Sanger Papers, SSC.
268 *But she did read an editorial*: Madeline Gray, *Margaret Sanger: A Biography of the Champion of Birth Control* (New York: Richard Marek Publishers, 1979), p. 435.
269 *"lust and animalistic mating"*: Ibid.
269 *"R.C. Church is getting more defiant"*: Ibid.
270 *"If a woman takes this medicine"*: John T. Noonan, *Contraception: A History of Its Treatment by the Catholic Theologians and Canonists* (Cambridge, MA: Harvard University Press, p. 461.
271 *"perfect health or perfect vision"*: Ibid., p. 463.

CHAPTER THIRTY

272 *"I thought it was so strange"*: Laura Pincus Bernard, interview conducted by the author, July 2013.
272 *"La señora de las pastillas"*: Ibid.
273 *"Many of them were never actually married"*: Ibid.
274 *medical team in Puerto Rico couldn't handle*: Transcript of Adaline Satterthwaite interview conducted by James Reed, June 1974, Schlesinger-Rockefeller Oral History Project.
274 *Haitian women to follow directions*: Laura V. Marks, *Sexual Chemistry* (New Haven, CT: Yale University Press, 2001), p. 104.
275 *diaphragms and jellies representing the next largest*: Robert Sheehan, "The Birth Control 'Pill'," *Fortune*, April 1958, p. 222.
275 *"It is no news"*: I. C. Winter to Gregory Pincus, December 29, 1958, Gregory Pincus Papers, LOC.

CHAPTER THIRTY-ONE

276 *"use any drug, medicinal article, or instrument"*: J. C. Ruppenthal, *Journal of the American Institute of Criminal Law and Criminology*, Vol. 10, August 1919, p. 53.

276 *"exhibit, sell, prescribe, provide"*: Fred Kaplan, *1959: The Year Everything Changed* (Hoboken, NJ: John Wiley & Sons, 2009), p. 226.

278 *"I don't know if you will approve"*: Margaret Marsh and Wanda Ronner, *The Fertility Doctor: John Rock and the Reproductive Revolution* (Baltimore, MD: Johns Hopkins University Press, 2008), pp. 213–14.

278 *"Enovid is an artificially made hormone"*: Ibid., p. 214.

279 *"pill might be the theological way out"*: Bernard Asbell, *The Pill: A Biography of the Drug That Changed the World* (New York: Random House, 1995), p. 154.

279 *"after a Chicago vice squad"*: Gay Talese, *Thy Neighbor's Wife* (New York: Doubleday, 1980), p. 126.

280 *a prescription for Enovid*: Katharine Dexter McCormick to Gregory Pincus, August 3, 1959, Gregory Pincus Papers, LOC.

280 *they might catch heat*: Paul Vaughan, *The Pill on Trial* (New York: Coward-McCann, Inc., 1970), p. 50.

281 *"numbers of Indians, Chinese,* et al.*"*: Robert Coughlan, "World Birth Control Challenge," *Life* magazine, November 23, 1959, p. 170.

CHAPTER THIRTY-TWO

283 *or maintain their professional education*: Subcommittee of the House Committee on Government Operations, Hearing on False and Misleading Advertising (Prescription Tranquilizing Drugs) 85 Cong. 2d (Washington, D.C., 1958), pp. 150, 226.

283 *father of ten children*: Loretta McLaughlin, *The Pill, John Rock, and the Church* (Boston: Little, Brown, 1982), p. 144.

284 *"everybody and her sister would be taking it"*: Ibid., pp. 143–44.

285 *"communist political and economic domination"*: President's Committee to Study the United States Military Assistance Program, *Letter to the President of the United States from the President's Committee to Study the United States Military Assistance Program and the Committee's Final Report* (Washington, D.C.), August 17, 1959, pp. 94–98.

285 *"I cannot imagine anything more emphatically"*: Matthew Connelly, *Fatal Misconception* (Cambridge, MA: Belknap Press of Harvard University Press, 2008), p. 187.

286 *couldn't prove that it* didn't *help arthritis*: McLaughlin, *The Pill, John Rock, and the Church*, p. 141.

286 *"We were in no hurry"*: Ibid., p. 142.

287 *"incomplete and inadequate"*: Pasquale DeFelice to William Crosson, September 25, 1959, John Rock Papers, CLM.

287 *"clearly outlined in our N.D.A."*: Margaret Marsh and Wanda Ronner, *The Fertility Doctor: John Rock and the Reproductive Revolution* (Baltimore, MD: Johns Hopkins University Press, 2008), p. 219.

288 *"the light of the obstetrical world"*: McLaughlin, *The Pill, John Rock, and the Church*, p. 143.

288 *"training you've had in female cancer"*: Ibid., p. 142.

288 *question seemed reasonable and important*: Ibid., p. 143.

289 *"I've only met about three doctors"*: Ibid.

289 *sixty-one obstetrician-gynecologists*: FDA memo, May 11, 1960, Gregory Pincus Papers, LOC.

290 *among his sixty-one experts*: Marsh and Ronner, *The Fertility Doctor*, p. 220.

290 *former joke writer for Groucho Marx*: Paul Vaughan, *The Pill on Trial* (New York: Coward-McCann, Inc., 1970), p. 52.

291 *adhesions of the labia and hypertrophy of the clitoris*: Ibid.

CHAPTER THIRTY-THREE

294 *three hundred thousand members*: James R. Petersen, *The Century of Sex: Playboy's History of the Sexual Revolution, 1900–1999* (New York: Grove Press, 1999), p. 264.

294 *results of a Gallup poll showed*: George Gallup, "Facts on Birth Control? A Loud 'Yes,'" *Ogden Standard-Examiner*, February 17, 1960, p. 3.

294 *massive protests on campus*: "Sympathy for Ousted Prof—But That's All," *Mt. Vernon Register-News*, April 12, 1960, p. 8.

295 *"Birth control and contraceptive practices"*: Margaret Sanger, "Population Planning," *New York Times*, January 3, 1960, p. E8.

295 *"neither birth control nor foreign aid"*: "Kennedy Renews Bid on Primaries," *New York Times*, January 4, 1960, p. 1.

296 *"You are young"*: Margaret Sanger to John F. Kennedy, January 11, 1960, Margaret Sanger Papers, SSC.

296 *He had begun testing . . . on men*: Gregory Pincus, *The Control of Fertility* (New York: Academic Press, 1965), p. 191.

296 *She had already committed $152,000*: Katharine Dexter McCormick to Margaret Sanger, January 2, 1960, Margaret Sanger Papers, SSC.

296 *doctors leading those experiments*: Seth S. King, "British Find Birth-Control Pills Cause Too Many Side Effects," *New York Times*, March 31, 1960, p. 41.

296 *"Don't be afraid of the number"*: "Pope, in Palm Sunday Homily, Makes Plea for Large Families," *New York Times*, April 11, 1960, p. 1.

297 *leaders in the Vatican were growing concerned*: John T. Noonan, *Con-*

traception: A History of Its Treatment by the Catholic Theologians and Canonists (Cambridge, MA: Harvard University Press, 1965), p. 490.

297 *theologians like Reinhold Niebuhr*: Ibid.

297 *"I am prepared to go to war"*: John Rock to William Crosson, April 18, 1960, John Rock Papers, CLM.

297 *At least two of the doctors*: Memo from W. H. Kessenich to FDA Commissioner George P. Larrick, May 11, 1960, John Rock Papers, CLM.

298 *"possible objections from some quarters"*: Ibid.

298 *"as far as they could tell"*: Ibid.

298 *DeFelice gave him the news*: William Crosson to Pasquale DeFelice, April 7, 1960, CLM.

299 *wrote and mailed a memo the same day*: Ibid.

299 *"Approval was based on the question of safety"*: "U.S. Approves Pill for Birth Control," *New York Times*, May 10, 1960.

299 *"interfere with the production of the ova"*: "Birth Control Pills Approved as Being Safe," *Denton Record-Chronicle*, May 16, 1960, p. 12.

299 *His daughter remembered no reaction*: Laura Pincus Bernard, interview conducted by the author, July 2013.

300 *"Well, the people, I would say"*: David M. Oshinsky, *Polio: An American Story* (Oxford, UK, and New York: Oxford University Press, 2005), p. 211.

301 *"Do not forget"*: "The New Pill for the Morning After," *Sydney Sun*, January 10, 1967.

302 *"weed out all the negative points"*: Elizabeth Siegel Watkins, *On the Pill* (Baltimore, MD: Johns Hopkins University Press, 1998), p. 36.

302 *"UNFETTERED"*: Ibid., photo insert.

303 *sales would increase 135 percent*: "G. D. Searle & Company," *International Directory of Company Histories*, 1996, *Encyclopedia.com*, http://www.encyclopedia.com/doc/1G2-2841600069.html (accessed December 12, 2013).

304 *"What I am busy over"*: Katharine Dexter McCormick to Margaret Sanger, June 15, 1960, Margaret Sanger Papers, SSC.

305 *more men undergo vasectomies*: Armond Fields, *Katharine Dexter McCormick: Pioneer for Women's Rights* (Westport, CT: Praeger, 2003), p. 296.

305 *struggling to kick her addiction*: Ellen Chesler, *Woman of Valor* (New York: Simon & Schuster, 1992), p. 458.

305 *"No one will really miss you"*: Barbara Benoit to Margaret Sanger, July 18, 1960, Margaret Sanger Papers, SSC.

305 *"rudderless drifting towards death"*: Dorothy Brush to Stuart Sanger, December 26, 1963, Margaret Sanger Papers, SSC.

305 *"so much government and other money"*: Margaret Sanger to Gregory Pincus, June 14, 1960, Gregory Pincus Papers, LOC.

306 *"The church hierarchy opposes"*: John Rock, "We Can End the Battle Over Birth Control," *Good Housekeeping*, July 1961, p. 44.

307 *One night before dinner*: Dr. Edward E. Wallach, interview conducted by the author, April 2013.

308 *pluck a pink flower*: Ibid.

308 *began to dance*: Ibid.

EPILOGUE

309 *since the exile of Adam and Eve*: Russell Shorto, "Contra-Contraception," *New York Times Magazine*, May 7, 2006, http://www.nytimes.com/2006/05/07/magazine/07contraception.html?pagewanted=all (accessed February 18, 2014).

309 *"central fact"*: Mary Eberstadt, *Adam and Eve After the Pill* (San Francisco, CA: Ignatius Press, 2012), p. 11.

309 *"solve the recurrent religious dispute"*: John Rock, *The Time Has Come* (London: The Catholic Book Club, 1963), unnumbered page.

310 *a majority of the committee members*: "Margaret Sanger is Dead at 82," *New York Times*, September 7, 1966, p. 1.

310 *"total" . . . a "special form"*: Robert McClory, *Turning Point* (New York: Crossroad, 1995), p. 139.

310 *"intrinsically dishonest"*: Ibid.

311 *G. D. Searle paid him twelve thousand dollars*: Margaret Marsh and Wanda Ronner, *The Fertility Doctor: John Rock and the Reproductive Revolution* (Baltimore, MD: Johns Hopkins University Press, 2008), p. 282.

311 *"It frequently occurs to me, gosh"*: "Doctor Rock's Magic Pill," *Esquire*, December 1983.

312 *Bone marrow cancer*: Dr. Robert Salomon to Gregory Pincus, September 3, 1963, Gregory Pincus Papers, LOC.

312 *"I am healthier than I have been"*: Dwight J. Ingle, *Gregory Goodwin Pincus: A Biographical Memoir* (Washington, D.C.: National Academy of Science, 1971), p. 238.

312 *"'kissed' your picture"*: Letter to Gregory Pincus, April 24, 1962, Gregory Pincus Papers, LOC.

312 *Gloria Steinem switched*: Gloria Steinem, "The Moral Disarmament of Betty Coed," *Esquire*, September 1962, p. 98.

313 *"housewives earning money as prostitutes"*: "The Pill: How It Is Affecting U.S. Morals, Family Life," *U.S. News & World Report*, July 11, 1966, http://www.pbs.org/wgbh/amex/pill/filmmore/ps_revolution.html (accessed February 18, 2014).

314 *"I would rather be asked for the pills"*: Ibid.

314 *"Mrs. Stanley McCormick"*: Dedication, Gregory Pincus, *The Control of Fertility* (New York: Academic Press, 1965).

314 *grown in value to about $25,000*: Leon Speroff, M.D., *A Good Man: Gregory Goodwin Pincus* (Portland, OR: Arnica Publishing, 2009), p. 271.

314 *every possible minute with his wife*: Laura Pincus Bernard, interview conducted by the author, July 2013.

315 *"This did wonders"*: Interview with David Wagner conducted by Patricia Gossel, January 1995; David Wagner Collection; Division of Science, Medicine, and Society; National Museum of American History, Smithsonian Institution; Washington, D.C.

316 *earning about $130,000*: Ibid.

317 *dressed in proper business attire*: Armond Fields, *Katharine Dexter McCormick: Pioneer for Women's Rights* (Westport, CT: Praeger, 2003), p. 297.

317 *"I knew I was right"*: Lloyd Shearer, "Margaret Sanger: Fifty Years of Crusading," *Parade Magazine*, December 1, 1963, p. 6.

318 *"willing to accept scorn and abuse"*: Esther Katz, ed., *The Selected Papers of Margaret Sanger*, Vol. 3 (Urbana: University of Illinois Press, 2010), p. 491.

318 *"population growth, when uncontrolled"*: Ibid.

319 *"vision was of a world"*: John Reedy, "Margaret Sanger, 1884–1966, R.I.P.," *Ave Maria*, September 24, 1966, pp. 5–6.

319 *"in a mere six years"*: "Freedom from Fear, *Time* magazine, April 7, 1967.

320 *The pill . . . lowered the cost of pursuing careers*: Claudia Goldin and Lawrence F. Katz, "The Power of the Pill: Oral Contraceptives and Women's Career and Marriage Decisions," *Journal of Political Economy* 110, no. 4 (2002), pp. 730–70.

320 *30 percent of the convergence*: Martha J. Bailey, Brad Hershein, and Amalia R. Miller, "The Opt-In Revolution? Contraception and the Gender Gap in Wages," May 13, 2012, http://www-personal.umich.edu/~baileymj/Opt_In_Revolution.pdf (accessed February 16, 2014).

321 *"first generation of oral contraceptives"*: Michelle Fay Cortez, "Birth-Control Pills Cut Cancer, Lengthen Women's Lives in Study," Bloomberg News, published March 11, 2010, http://www.bloomberg.com/apps/news?pid=newsarchive&sid=amLgSVxKl4zw (accessed February 16, 2014).

Selected Bibliography

Allyn, David. *Make Love, Not War.* Boston: Little, Brown, 2000.

Asbell, Bernard. *The Pill: A Biography of the Drug That Changed the World.* New York: Random House, 1995.

Bailey, Beth. *Sex in the Heartland.* Cambridge, MA: Harvard University Press, 1999.

Baker, Jean H. *Margaret Sanger: A Life of Passion.* New York: Hill and Wang, 2011.

Briggs, Laura, *Reproducing Empire.* Berkeley: University of California Press, 2008.

Brynner, Rock, and Trent Stephens. *Dark Remedy.* New York: Basic Books, 2001.

Callaway, Enoch. *Asylum: A Mid-Century Madhouse and Its Lessons about Our Mentally Ill Today.* Westport, CT: Praeger, 2007.

Carpenter, Daniel. *Reputation and Power.* Princeton, NJ: Princeton University Press, 2010.

Chesler, Ellen. *Woman of Valor: Margaret Sanger and the Birth Control Movement in America.* New York: Simon & Schuster, 2007.

Connelly, Matthew. *Fatal Misconception: The Struggle to Control World Population.* Cambridge, MA: Belknap Press of Harvard University Press, 2008.

Critchlow, Donald T. *Intended Consequences.* Oxford: Oxford University Press, 1999.

Diamond, Jared. *Why Is Sex Fun?* New York: Basic Books, 1997.

Dietz, James L. *Economic History of Puerto Rico.* Princeton, NJ: Princeton University Press, 1986.

Djerassi, Carl. *This Man's Pill: Reflections on the 50th Birthday of the Pill.* Oxford: Oxford University Press, 2001.

Eberstadt, Mary. *Adam and Eve After the Pill.* San Francisco, CA: Ignatius Press, 2012.

Engelman, Peter C. *A History of the Birth Control Movement in America.* Santa Barbara, CA: Praeger, 2011.

Escoffier, Jeffrey. *Sexual Revolution.* New York: Thunder's Mouth Press, 2003.

Fields, Armond. *Katharine Dexter McCormick: Pioneer for Women's Rights.* Westport, CT: Praeger, 2003.

Foucault, Michel. *The History of Sexuality,* Vol. 1. New York: Vintage Books, 1990.

Gordon, Linda. *The Moral Property of Women: A History of Birth Control Politics in America.* Urbana: University of Illinois Press, 2007.

Grant, Linda. *Sexing the Millennium.* New York: Grove Press, 1994.

Gray, Madeline. *Margaret Sanger: A Biography of the Champion of Birth Control.* New York: Richard Marek Publishers, 1979.

Halberstam, David. *The Fifties.* New York: Villard Books, 1993.

Harvey, Brett. *The Fifties: A Women's Oral History.* New York: HarperCollins, 1993.

Hatt, Paul K. *Backgrounds of Human Fertility in Puerto Rico.* Princeton, NJ: Princeton University Press, 1952.

Hill, Reuben, et al. *The Family and Population Control.* Chapel Hill: University of North Carolina Press, 1959.

Hilts, Philip J. *Protecting America's Health.* Chapel Hill: University of North Carolina Press, 2003.

Katz, Esther, ed. *The Selected Papers of Margaret Sanger,* Vol. 3. Urbana: University of Illinois Press, 2010.

Kennedy, David M. *Birth Control in America.* New Haven: Yale University Press, 1970.

Lader, Lawrence. *The Margaret Sanger Story.* New York: Doubleday, 1955.

Lepore, Jill. *The Mansion of Happiness: A History of Life and Death.* New York: Vintage Books, 2012.

Lerner, Max. *America as a Civilization.* New York: Simon & Schuster, 1957.

Maisel, Albert Q. *The Hormone Quest.* New York: Random House, 1965.

Marks, Laura V. *Sexual Chemistry.* New Haven, CT: Yale University Press, 2001.

Marsh, Margaret, and Wanda Ronner. *The Fertility Doctor: John Rock and the Reproductive Revolution.* Baltimore, MD: Johns Hopkins University Press, 2008.

May, Elaine Tyler. *America and the Pill.* New York: Basic Books, 2010.

McLaughlin, Loretta. *The Pill, John Rock, and the Church: The Biography of a Revolution.* Boston: Little, Brown, 1982.

Metalious, Grace. *Peyton Place.* Boston: Northeastern University Press, 1999.

Meyerowitz, Joanne, ed. *Not June Cleaver: Women and Gender in Postwar America, 1945–1960.* Philadelphia: Temple University Press, 1994.

Noonan, John T., Jr. *Contraception: A History of Its Treatment by the Catholic Theologians and Canonists.* Cambridge, MA: Harvard University Press, 1965.

Oshinsky, David M. *Polio: An American Story*. Oxford: Oxford University Press, 2005.

Oudshoorn, Nelly. *Beyond the Natural Body: An Archaeology of Sex Hormones*. London: Routledge, 1994.

Petersen, James R. *The Century of Sex: Playboy's History of the Sexual Revolution, 1900–1999*. New York: Grove Press, 1999.

Pincus, Gregory. *The Control of Fertility*. New York: Academic Press, 1965.

———. *The Eggs of Mammals*. New York: Macmillan, 1936.

Ramirez de Arellano, Annette B. *Colonialism, Catholicism, and Contraception*. Chapel Hill: University of North Carolina Press, 1983.

Reed, James. *From Private Vice to Public Virtue: The Birth Control Movement and American Society Since 1830*. New York: Basic Books, 1978.

Roach, Mary. *Bonk*. New York: W. W. Norton & Company, 2008.

Rock, John. *The Time Has Come*. London: Catholic Book Club, 1963.

Sanger, Margaret. *The Autobiography of Margaret Sanger*. Mineola, NY: Dover, 1971.

Smith, Janet E. *Humanae Vitae: A Generation Later*. Washington, D.C.: Catholic University of America Press, 1991.

Speroff, Leon. *A Good Man: Gregory Goodwin Pincus*. Portland: Arnica Publishing, 2009.

Stycos, J. Mayone. *Family and Fertility in Puerto Rico*. New York: Columbia University Press, 1955.

Suitters, Beryl. *Be Brave and Angry: Chronicles of the International Planned Parenthood Federation*. London: International Planned Parenthood Federation, 1973.

Talese, Gay. *Thy Neighbor's Wife*. New York: Doubleday, 1980.

Tentler, Leslie Woodcock. *Catholics and Contraception: An American History*. Ithaca, NY: Cornell University Press, 2004.

Tone, Andrea. *Devices and Desires: A History of Contraceptives in America*. New York: Hill and Wang, 2001.

———, ed. *Controlling Reproduction*. Wilmington, DE: Scholarly Resources, 1997.

Vaughan, Paul. *The Pill on Trial*. New York: Coward-McCann, 1970.

Villee, Claude A., ed. *Control of Ovulation*. Oxford: Pergamon Press, 1961.

Watkins, Elizabeth Siegel. *On the Pill: A Social History of Oral Contraceptives, 1950–1970*. Baltimore, MD: Johns Hopkins University Press, 1998.

Wood, Clive, and Beryl Suitters. *The Fight for Acceptance: A History of Contraception*. Aylesbury, UK: Medical and Technical Publishing Co., 1970.

Zuckerman, Solly. *Beyond the Ivory Tower*. London: Scientific Book Club, 1970.

Index

HEROES

Contents

Wilma Rudolph

Written by Kerrie Ann Capobianco
Illustrated by Marjorie Scott

1940

Wilma Rudolph was born
in St. Bethlehem, Tennessee.

1944

When Wilma was four
she got polio.
Polio made Wilma's legs weak.
It was hard for Wilma
to use her left leg.
It was hard for Wilma to walk.

Wilma's mother
took her to the doctor.
The doctor said,
"Rub the leg every day!"
Wilma's mother and brothers
and sisters rubbed her leg
four times a day.
The rubbing helped Wilma's leg.

When Wilma was five,
the doctor put a brace
on her leg.
The brace helped Wilma walk.
Wilma had to wear the brace
all the time.

1951

How long did Wilma wear her leg brace?

One day, Wilma took the brace off her leg.
She went to church without the brace.
She could walk without the brace.

1954

Wilma liked to play sports.
She liked to play basketball.

Wilma liked to run.
She liked to run in races.
She went to track meets.
At her first track meet,
Wilma lost all the races.
Wilma needed a coach to help her.
A man named Ed Temple
was a good coach.
He coached Wilma.
Wilma trained hard.

1958

Wilma went to
Tennessee State University.
She worked hard.
She trained hard.
She ran races.
She did part-time work
so she could stay
in school.

1960

Wilma went
to the Olympic Games
in Rome.
She ran in
the 100-meter race.
She ran in
the 200-meter race.
And she ran in
the 400-meter relay race.
Wilma won
all her races.
She won
three gold medals.
She was the first
American woman
to win three gold medals.

Daily Reporter

November 1994

Wilma Rudolph died today.

She did what people said was too hard. She beat the polio that she had. She walked. She ran. She raced. She won! Wilma Rudolph was a hero.

60
65

Bubbles

Written by Janet Slater Bottin
Illustrated by Phyllis Pollema-Cahill

Bubbles woke up.
She could hear strange sounds.
She looked up at the bed.
Melissa was asleep.
Bubbles looked after Melissa.
Bubbles helped Melissa hear things.

Melissa could not hear sounds.
She could not hear low sounds.
She could not hear high sounds.
She could not hear the strange sounds
that Bubbles could hear.
Bubbles lay down again.
She put her head on her paws.

Then she heard the strange sound again!
It was a bad sound!
The sound came from
the twins' bedroom!
Bubbles didn't bark.
Melissa would not hear her
if she barked.
Bubbles jumped up onto Melissa's bed.
Then she jumped down
onto the floor again.
Up... down... up... down... up... down.

What are some strange sounds that you might hear in the night?

Melissa woke up. She sat up.
"What do you want, Bubbles?
Do you want to go out?"
she said.

Melissa got out of bed.
She went to the door.
Bubbles ran
to the twins' bedroom!
Melissa said,
"What is it, Bubbles?"
Bubbles jumped
at the door
of the twins'
bedroom.

What do you think Bubbles has heard?

Melissa ran to the door and put on the bedroom light.
Bubbles ran to little Zeke's bed.
Bubbles put her paws on his bed.
"ZEKE!" cried Melissa.
Her little boy was blue.
He was not breathing!
Melissa ran to Zeke.
She picked him up.
She held him upside down.
She patted his back.
She called for help.
Bubbles looked on.

What kind of things would you need to do to train a dog like Bubbles?

Bubbles heard the ambulance.
She saw people come to help.
She saw Melissa and little Zeke
get into the ambulance.
She heard the ambulance drive away.

She had helped her friend Melissa.
She had saved little Zeke.

Daily Reporter

Hearing Dog, Bubbles, a Hero

Bubbles, a ten-year-old hearing dog, saved a child today. Bubbles woke up her owner, Melissa, when she heard a strange sound. Zeke, one of Melissa's twins, had stopped breathing.

This is not the first time that Bubbles has helped Melissa. One time, Bubbles saved Melissa when a log fell out of the fire.

Another time, Bubbles saved Melissa when she had a crash in her car.

Bubbles is a real hero!

Brent Meldrum

Written by
Sharon Capobianco
Illustrated by Jeannie Ferguson

From a true story:

Brent Meldrum and his friend
Tanya Branden were playing.
Tanya was eating something.
All of a sudden, Tanya began to choke.
Something was stuck
in her throat.
Tanya could not breathe.
She went blue.

Brent knew that
he had to help Tanya.
He had to stop
his friend from choking,
or she might die.
He had seen a person
choking on television.
He had seen what people did
to save the person's life.
He was going to do that
for Tanya.

Always be careful eating things that could stick in your throat!

Brent went behind Tanya.
He put his arms around her.
He put his hands together.
He pushed his thumbs up
under Tanya's ribs.
He lifted Tanya up,
off the ground.
Then he put her
back down again.
Tanya coughed up
what she was eating.
Brent had saved Tanya's life.

How do you think Brent felt about saving Tanya's life?

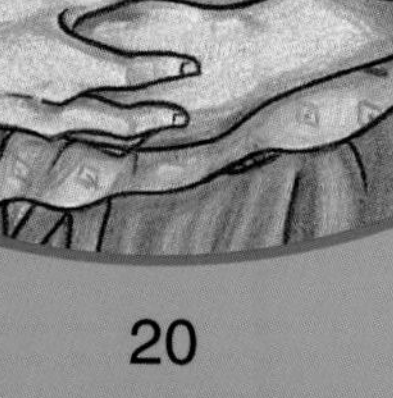

IMPORTANT!
Read this!

Brent Meldrum is a true hero!
But be careful!
Don't try what Brent did
on your friends!
It could hurt them!
Use it for a real emergency.

In this emergency,
Brent forced air
out of Tanya's lungs.
The air blew out food
that was stuck in her windpipe.
You could learn this method
with a parent or caregiver.
Like Brent Meldrum,
you could save a life!

A Five-year-old Hero

Today, a five-year-old boy saved the life of his friend. His friend Tanya was choking on something. Brent Meldrum used a method that helped his friend cough up what she was eating. A man named Henry Heimlich made up the method in 1974.

Reporter

August 6, 1986

Saves His Pal's Life

Heimlich showed people how to get out food that was stuck. Before he did this, lots of people had died from choking on food. We asked Brent to show us what he had done. Brent said he had seen the "Time-Life Remover" on TV. No matter what you call the method, it saved his pal's life.

Written by Louise Fenwick Illustrated by Mrinal Mitra

Mother Cat was a street cat.
She was a very smart street cat.
Mother Cat had five little kittens.

At night, Mother Cat went out to find food.
"I will be back soon," she said to her kittens.

One night, Mother Cat
got the food and set out for home.
When she got near her home,
Mother Cat heard a strange sound.
She saw tall flames jump up into the sky.
Her home was on fire!

What are some things that street cats have to learn about?

Mother Cat ran into her home.
She ran into the flames.
She ran to the place
where her little kittens were.
"I will get you out,"
she said to her kittens.

How do you think Mother Cat was feeling?

Mother Cat got the first little kitten
by the back of the neck.
She ran into the flames and out on to the street.
"Stay here," she said.
"I will come back with your brothers and sisters."

Mother Cat went back into the flames.
She took the second little kitten out.

Then she took the third little kitten out.

And then she took the fourth little kitten
out from the fire and on to the street.

Then she ran back into her home
for the last time.

The fifth little kitten lay still.
Its fur was hot.

"Oh no!" cried Mother Cat.
She picked up the little kitten.
She ran into the flames and out on to the street.

She put the little kitten down by her other kittens.
She began to lick its hot fur.

A firefighter came over.
He sat down by Mother Cat.
He looked at the kittens, one by one.

"It's all right," he said to Mother Cat.
"All your kittens are safe."
The firefighter put Mother Cat and her kittens in a box.
He took them to the vet.

"You are a very brave cat," said the vet.
She washed Mother Cat's fur.
She rubbed some cream on Mother Cat's burns.

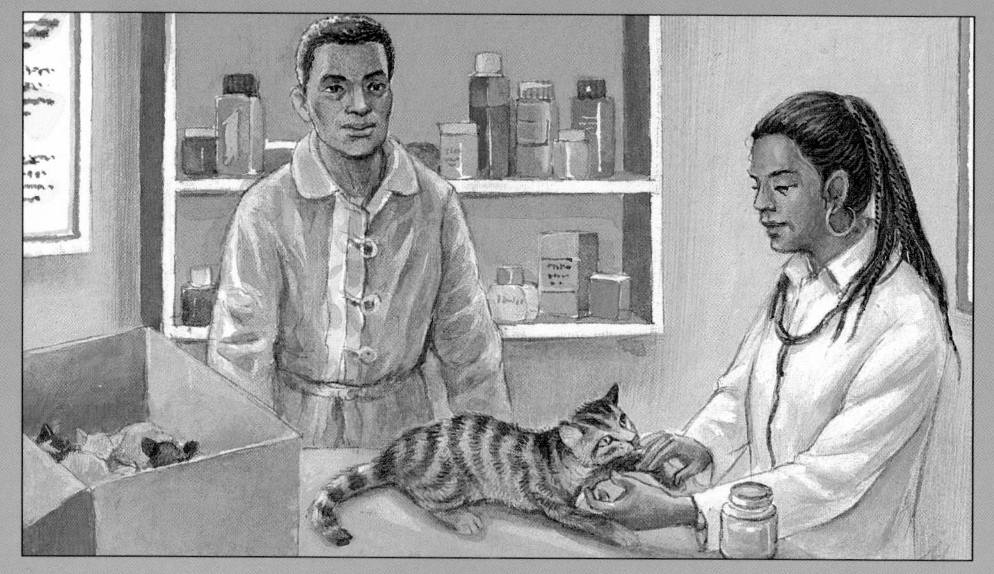

The firefighter took Mother Cat
and her five little kittens to live with him.
He called the kittens
Sooty, Cinders, Flame, Firefly, and Smokey.
Mother Cat did not have to go out at night
to find food again.

What event changed Mother Cat's life?

Glossary

- **brace** – a thing made to support or rest a part of the body that is weak or needs to be held in the right place
- **coach** – a person who trains people for events or helps them learn the skills they need
- **emergency** – an event or accident that needs urgent action
- **hearing dog** – a dog that will alert a person with a hearing disability to sounds such as a telephone or alarm
- **heroes** – people who have shown bravery and courage and who are looked up to by those around them
- **polio** – a disease that can cause paralysis of parts of the body
- **street cat** – a homeless cat
- **track meets** – a series of events for which athletes come together to compete